PROSTHETIC IMMORTALITIES

CARY WOLFE, SERIES EDITOR

71 *Prosthetic Immortalities: Biology, Transhumanism, and the Search for Indefinite Life*
Adam R. Rosenthal

70 *The Memory of the World: Deep Time, Animality, and Eschatology*
Ted Toadvine

69 *Hermes I: Communication*
Michel Serres

68 *Nietzsche's Posthumanism*
Edgar Landgraf

67 *Subsurface*
Karen Pinkus

66 *Making Sense in Common: A Reading of Whitehead in Times of Collapse*
Isabelle Stengers

65 *Our Grateful Dead: Stories of Those Left Behind*
Vinciane Despret

64 *Prosthesis*
David Wills

63 *Molecular Capture: The Animation of Biology*
Adam Nocek

62 *Clang*
Jacques Derrida

61 *Radioactive Ghosts*
Gabriele Schwab

60 *Gaian Systems: Lynn Margulis, Neocybernetics, and the End of the Anthropocene*
Bruce Clarke

59 *The Probiotic Planet: Using Life to Manage Life*
Jamie Lorimer

(continued on page 323)

PROSTHETIC IMMORTALITIES

BIOLOGY, TRANSHUMANISM, AND THE SEARCH FOR INDEFINITE LIFE

ADAM R. ROSENTHAL

FOREWORD BY DAVID WILLS

posthumanities 71

University of Minnesota Press
Minneapolis
London

Published by the University of Minnesota Press
111 Third Avenue South, Suite 290
Minneapolis, MN 55401-2520
http://www.upress.umn.edu

ISBN 978-1-5179-1657-2 (hc)
ISBN 978-1-5179-1658-9 (pb)

A Cataloging-in-Publication record for this book is available from the Library of Congress.

Printed on acid-free paper

CONTENTS

FOREWORD

Preferring Being Condemned Not To

DAVID WILLS

So far, we animals all die.

When we die, one by one, we will have come to know that we all die, even if "knowing" it, really knowing it, escapes our consciousness. But before we die, we know—more or less consciously, depending on our ontological status—that what we will die of, that is, what will bring us to the moment of dying—is not knowing *when* we will die. For not knowing when we will die constitutes our mortality and our finitude; in a sense, therefore, it is what we die of. The fact of our having to die, but not knowing when that will occur, is what makes death an issue for us; it is what brings us agonizingly all the way to our final agony, which is the moment and experience of death. Thus, in the moments leading right up to the point where we are becoming dead, our nonknowledge of death is what ails us. The sequential proximity of not knowing the "when" of dying is so close to the when itself of dying itself that this sequence of events functions as a type of *consequential* relation. It makes our mortal nonknowledge of death—beginning with our nonknowledge about when it will occur—something of our cause of death, what we die of. Death will creep up on us, one way or another, and before we know it, we will have died; but in effect, even as we die, we will not know—going toward the white light notwithstanding—*when* we will have died, and that ignorance is, according to my argument, the cause of our death. But it is also our cause when it comes to death; it is what makes death an issue—our issue, our cause, our all-consuming cause or *causa,* our thing.

Not knowing when we have died, but dying all the same; knowing we die even without knowing that we have died when we do die, or when we have died—that is the form of our finitude, what makes death our cause. It binds us to a death that we know of but that we will never know. It is what *condemns us to die.* Following that logic, if mortality means knowing we will die but not knowing when, then immortality may mean less not dying than it may mean knowing dying itself, knowing the moment we die, experiencing it in or as time. As long as we remain mortal, we will not know dying in that sense; were we to become immortal, or so my argument goes, we would conversely know dying and so *do dying.* That would be one way to understand Blanchot's "dead—immortal," as described by the first-person narrator who is given a reprieve in *The Instant of My Death.*[1] Immortality would be really something if it could be the something of knowing the transition from life to death as an experience of that sort of oppositional difference, but knowing it precisely as something other than a binary oppositional difference, no longer a knowledge opposed to nonknowledge, no longer a death simply opposed to life; if it could be knowing the meaning of life as distinct from the meaning of death in the sense of knowing death differently from how we know, so far, that we are condemned to die. In that way it would be something other than being condemned not to die.

The idea that in being mortal we are condemned to die is underlined by Jacques Derrida in his seminars on *The Death Penalty.* He makes the point in order to distinguish between us everyday ordinary mortals and those relatively few, exceptionally victimized mortals, who are subject to a decision made by this or that state power—exercising its sovereign right to the exception—and so are condemned to death: "The mortal that I am knows he is condemned to die, but . . . does not know the moment, the date, the precise hour that he will die. . . . The concept of the death penalty supposes that the state, judges, society, the bourreaux and executioners, that is, third parties, have mastery over the time of life of the condemned one."[2] In order to perform that exceptional political condemnation, that cruel and arbitrary imposition of mortality, the state mobilizes a whole penal technology of death, what might be called a whole "penal colonization" of death. For example, in a particularly egregiously penalizing state of the global north, the United States, the death penalty begins to take place by a progressive colonizing, by means of imprisonment and often solitary confinement, of the remaining time of life of one who is condemned, stretching it out toward death over a period of months, years, decades. That is how the U.S. death-penalty industry begins to prepare to put to death. Consider this instance: in fall 2015, I began writing a book on the temporal technology of the death penalty in which I referred to Richard Glossip, who was sentenced to death for murder in Oklahoma in 1998, and who had, by the time I finished the book in September 2018, eaten his final meal three times. That was then. Now, eight years later, as I begin to write this foreword, twenty-five years since Glossip's sentencing, he appeared on April 26, 2023, at a hearing before the Oklahoma Pardon and Parole Board, which, by a split 2–2 vote, declined his petition for clemency and reaffirmed the scheduling of what is now his ninth assigned execution date, for May 18. That was despite an argument in Glossip's favor by none less than the current Oklahoma attorney general, Gentner Drummond, who has serious doubts about his guilt. On May 6, 2023, however, the Supreme Court stayed Glossip's execution pending a review of his petitions for appeal, and he survives languishing still.[3]

By interrupting the flow of mortal time—the time of a natural life experienced as uninterrupted until, as if out of nowhere, death arrives—and by introducing into that flow a contrived moment at which a condemned person's life will end, the death penalty emphasizes the technological force of chronology, the idea that what time is really for is less to count out our minutes, hours, and days than to mark and inscribe our final moment, to determine the instant of our death. The death penalty serves to expose what I call that technicity or prostheticity of mortal time. As I argued in *Killing Times,* "From the moment we understood ourselves to be mortal, we became subject to a type of countdown that functioned as a putatively external system of control over our lives; we adopted, and adapted to time as that prosthesis, even if we claimed or pretended to be able to avoid or outwit it."[4] Once there is mortality, there is that prosthesis called time; but once there is the death penalty, it is as if the prosthesis called time has taken on a different mechanicity, with the consequence that we are all prevented from experiencing prosthetic time as though it were natural, however incommensurably different the experience of such nonnaturality, and of the threat of death, might be for most of us who are merely condemned to die, in contrast to those relatively few, but far too many, who wait on death row, having been condemned to death and having had a moment of death assigned to them.

Hegel develops, no doubt in spite of himself, a concept of prosthetic time that depends on blood, on a lifeblood that is able, uncannily, to operate both inside and outside the body. "Universal blood" is referred to briefly in *The Phenomenology of Spirit* as synonymous with "the absolute concept . . . the simple essence of life, the soul of the world."[5] This blood is the flow of a spirit that has "annulled" time in absolute knowledge; it is the infinite flux of an "unresting and unhalting time" that has collapsed into itself.[6] In *The Philosophy of Nature,* where animal blood flowing inside a body is explicitly in question, that flux of unhalting time would seem to be attainable only once an individual dies, allowing a "more beautiful Nature, spirit" to emerge from the dead husk of the body.[7] Before that, blood is the vehicle of organic irritability, "inward activity, pulsation, living self-movement, the material of which can be only a fluid, the living blood."[8] Indeed, animal blood is "the individual life itself."[9] On closer examination, however, that animal blood is seen to be involved in the same sort of dialectical sublation that has led from sense-certainty all the way to spirit; it is simultaneously "the ground and the movement . . . [a] stepping aside of itself . . . and the supersession *[das Aufheben]* of its standing aside," allowing it to become an "endless, unbroken unrest of welling forth."[10] The fluidity of the blood is therefore what raises the organism "into pure ideality, perfectly transparent universality."[11]

Effectively, then, animal blood pulsating through the body rejoins the universal blood of the absolute concept, immortalizing itself as it were. In fact, there is a case where blood steps aside of itself, even steps outside of itself, in a continuous manner; that is the pulmonary system, which, as Hegel describes it, operates less by sucking air into the lungs than by having the blood step out to meet and conquer the air: "the pulmonary system . . . [is] the true, ideal process with the outer world of inorganic Nature, with the Element of air."[12] In this way, blood achieves "sheer victory" over the immaterial and inorganic, allowing the organism to attain "the direct melting of the nonorganic into organic fluidity."[13] Blood in Hegel is thus the means by which the body reaches outside itself to restore, complete, or fulfill itself, to quench its thirst. It is therefore the body's first and best prosthesis, its ideal assimilator of inorganic otherness, a perfect sublation machine that functions thanks to its "absolute thirst, its unrest within itself and against itself."[14] The thirst it has is a thirst for life in the form of victory over the inorganic and satisfaction of bodily needs, but it is also a thirst for victory over time. It is a synchronization of, on the one hand, internal prosthetic time, the pulsating animal time of "axially rotating" blood ("self-pursuing movement, this absolute interior vibration . . . the individual life of the whole in which there is no distinction—animal time"[15]), and on the other hand of seamless external prosthetic time, that of universal blood, idealized and spiritized, a time so unresting and unhalting that its clunky prostheticity will have dissolved or melted into thin air, allowing the animal to reform as immortal spirit.

Hegel's concept of time as infinite flux, of time annulled by spirit, supposedly functions at a remove from any human or animal desire for immortality while at the same time reintroducing such a desire. Rosenthal's *Prosthetic Immortalities* makes it clear that any contemporary yearning for or promise of indefinite life will never easily emancipate itself from idealist or ontotheological thinking. His principal figure isn't Hegel but rather Plato, Descartes, or Heidegger; nor is his vehicle blood. However, his argument is that throughout that philosophical tradition, and all the way to current transhumanisms, biotechnologies, and life extensions, immortality will have always functioned as a prosthesis. A "general logic of prosthetic immortalization" will not just describe our phantasmatic attachment to immortality as a type of sinecure—imagine how much better life would be if we could avoid dying, if we could have the power of the gods!—but also delineate immortality as a structural feature of mortality's necessary imagining of its own outside: "Immortality does not simply follow life, but invents it, transforming that which will have been valued of life, interpreted as life, delimited as life, translated from life, in advance."

For Rosenthal, conceiving immortality requires thinking back all the way to, or through, what Derrida calls trace or *différance,* a type of drive that inhabits every *ipse* or *autos,* every self-thesis or self-positioning inasmuch as it seeks to extend itself beyond itself (as pro[s]-thesis), from the unknown or unknowable beginning of life to the unknown or unknowable end of life. Precisely as Rosenthal also suggests, in discussing Heidegger's being-toward-death—which is where he picks up on Derrida's emphasis on our being condemned to die—the prosthetic immortalizing that can be understood to impel life at its beginning functions as a through line governing putative future extensions "beyond" death. The whole problematic of current approaches to the question is therefore put to the test of Derrida's understanding of trace, and of a redefinition of life and finitude as survival, or surviving: "only a living being as finite being can have a future," according to Derrida, and that is thanks to a "relation to incalculability and undecidability as to the instant of my death."[16] One might understand it in the following way, via a sort of Derridean parsing of Heidegger: I come to life by means of the finite *différance,* or perhaps differenciation, of a lost, infinitely lost, origin; once I realize that I am alive, that I have been torn from that lost origin (or, in Hegelian terms, dropped out of the flux of unhalting time), I understand that I am mortal, living in a structure of respite, reprieve, suspension, or survival with respect to an anticipated end; my conception and organization of life depends as much on the idea of suspension or survival as it does on the idea of ending; that is what can be called my future, and it is obviously not mine but rather something deferred to me by another.

How then do current approaches to the question of immortality stand that test? In *Prosthetic Immortalities,* the analysis is pursued on two tracks: regarding an ontological relation to (im)mortality that works through the experience of "living as dying," elements or versions of which have just been outlined, and regarding a "material" or biological immortality that staves off physical decay and destruction. Biological immortalities divide in turn between, on the one hand, the existence of life-forms, such as bacteria or viruses (and indeed genes), which avoid death by replicating themselves or subdividing, and on the other hand, the possibility for humans themselves to eliminate death by dealing with it as something to be corrected or treated—for example, by modifying our genetic makeup in a way that will provide accession to some form of indefinite life or indefinite aging. Then, differently again, there are those who ask whether immortality need rely on "the forms of continuity, identity, and individuality familiar to us," whether we couldn't or shouldn't rather upload or "emulate" our brains to continue our existence. Rosenthal finds in the end that all of those approaches face conceptual challenges, as they were of their own making. Chief among examples of biological immortality is unicellular life, "the founding instance of all biological continuity . . . preserved in both the simplest and most complex organisms," but to speak of continuity in relation to cell life is to import a concept that derives from human or animal existence, and from our own problematic sense of ipseity. If we consider instead the metabolic operations of multicellular systems, we approach "something like the essence of the living," but in Rosenthal's reading, for example of Hans Jonas, that essence is nevertheless revealed as a prostheticity in the heart of autopoietic autonomy.

At the end of the line of prosthetic immortalities is cloning, represented by Rosenthal as a word in search of a concept, a fictive practice in search of an identity, which, now being double, it cannot possess. Cloning thus reveals the corruption and self-deception of every dream of immortality. It means that the book you are about to read will end on a question—am I, then, immortal?—a question that enfolds all the other inquiries it will have already pursued—especially, once more, whether and how I am condemned to die.

The fundamental distinction, underscored by the death penalty, between being condemned to die and condemned to death is allowed, by the end of Derrida's seminar, to extend to the "many techno-scientifico-capitalist mechanisms for distributing, in a terribly unequal way, the right to life, to longevity, and which therefore not only condemn to death a calculable number of individuals but also condemn to die prematurely an incalculable number of living beings, human and nonhuman."[17] We thus slide outside the purely judicial matter of imposing a time for putting to death and into the economicoeugenic project of letting die. The question of immortalization, of whether we can live longer, necessarily intersects with questions of who lives, for how long, and how well. To live longer is posited on the presumption that it is to live better—that most is best. Yet as long as prosthetic immortalizations consistently bury questions of quality of life under a quantitative mass of more life, they remain corrupted by eugenicist phantasms.

Rosenthal discusses the insistence on, or preference for, mortality of a Martin Hägglund or a Todd May, both of whom advocate the fragility and vulnerability it implies, in apposition, and perhaps opposition to the transhumanist confidence expressed by Nick Bostrom in the ability of a superintelligence to solve any problem: "disease, poverty, environmental destruction, unnecessary sufferings of all kinds."[18] Similarly, Yuval Noah Harari suggests that any ethical qualms we might have about potential abuses of such technologies should be offset by the ethical imperative that derives from their positive potential for "prolonging human life indefinitely, conquering incurable diseases, and upgrading our cognitive and emotional abilities."[19] Of course, this contrasting ethical assessment of technology was never limited to so-called new technologies, and that not because every technology is new but precisely because no technology is new—that is to say, technology is less a function of invention than of a *differantial* prosthetization always already in process; the human was never nontechnological. For the same reason, ethical debate about the adverse or favorable effects of a given technology is always headed for the same impasse: such effects cannot be forethought by this optimism or forestalled by that pessimism inasmuch as they are always already in effect in us.

Now, this means neither that every technological emergence is desirable or inevitable nor that a given use of a given technology cannot or should not be resisted or refused. The application of the death penalty involves modifying technology, in contradistinction to the use of lifesaving prostheses; refinements of the instant of death from ax or guillotine to lethal injection are no less perverse for presuming to improve the performance of an execution; and then there are the supersmart death-dealing bombs and drones hell-bent on destruction across various parts of the globe. *Prosthetic Immortalities* was written before October 7, 2023, before the murderous Hamas attack, and before the IDF set into operation—financed and supplied, and, it is now reported, even advised by the United States[20]—its scarcely discriminate pommeling and pulverizing, choking, and throttling death machines across Gaza. As of this date, some 23,000 people, including more than 10,000 children, with thousands more trapped or rotting under the rubble, have been forced to walk the knife-edge between being put to death and let die during this, our unspeakable present moment. So when it comes to the immortalities discussed here, we should be automatically suspicious of global north economizations that prioritize human extensions of life suddenly weaned from extensions of life expectancy such as are desired in the global south. We should be suspicious of a superintelligence the economics of which will—in the current international neoliberal market—determine that its ability to conquer disease is used selectively and sold expensively where it is needed most. We should be suspicious because we watched the international production and merchandising of Covid-19 vaccinations at the expense of our most exposed fellow humans. We should have no illusions about the marketing of immortalizing phantasms because we have already seen what drives the race to their realization. We have seen that in Jeffrey Epstein's reported dream of preserving intellectually supreme spermatozoa, in Peter Thiel's idealized antipodean bunker, and in Israel's right to self-preservation that comes at the expense of even the most meager level of Palestinian subsistence.

So far, though, we all die; we are all condemned to die.

As long as we live, we live in the time of an imminence that is also a reprieve or an abeyance. In *The Instant of My Death,* that reprieve was lived by Blanchot's narrator as "a feeling of extraordinary lightness, a sort of beatitude," such that in being "dead—immortal," he was perhaps experiencing "ecstasy."[21] If the transition from being condemned to death to being condemned to die induces ecstasy, we can imagine, and hope for similarly redemptive relief for those currently living bare lives, were we to open for them the passage from being condemned to live badly, condemned to being let die, to "simply" being condemned to die. From there, presuming we could get there, presuming the prioritizations and hierarchizations of technological possibility and accessibility were realigned in favor of the least fortunate, we could then begin to fantasize about the sort of ecstasy that might be in store for us, were we to be emancipated from our condemnation and made immortal in the ways imagined below.

But we are still some way from there. In the meantime, we are in time. We wait to die, at which point we will, in a sense, no longer be in time, which is precisely what will prevent us from having experiential knowledge of death. Outside that, in our phantasms of a still fictive future, we wait not to have to die. Were we to reach and realize that future, were we not to have to die, we might arrive at knowing dying itself, experiencing it as time, as I suggested to begin. Or else, perhaps, reaching and realizing that future, substituting for the prosthesis called time this or that prosthetic immortality, we might indeed emancipate ourselves from time so as to live forever in a forever that has no time. Something in this book whispers to me—many things among its many things in fact whisper—that in correcting the deficiency that is our mortality by means of a prosthetic immortality, we will not have learned anything; nor will we have experienced anything that was not already implied when we compared ourselves to the gods and began to desire to emulate them. We will not have learned anything, nor experienced anything in life or death or immortality until—no longer thinking like a human, no longer being human, no longer being at all—we assume a different relation of death to life, within which we are no longer simply seeking the ideal add-on, the perfect accessory to mortality, but instead coming to know death's relation to life to the extent of designing and constructing it, refining and replacing it, imitating and parodying it—not because we will have attained a supremely ingenious voluntarist intelligence but rather because we will have come to inhabit the very fissure and fissuring where life contracts with, and as, other than life.

PREFACE AND ACKNOWLEDGMENTS

Deconstruction's Last Word, or The Afterlife of Immortality

> Neither death nor immortal life can ever give anything, only a singular *surviving* can give. This is the *element* of this problematic.
>
> —Jacques Derrida, *Given Time*

In what is known as the final interview, the last, or nearly last, public testament put forward by French philosopher Jacques Derrida, amid reflections on the end of his life and the relationship between this each-time-unique life and life itself, the then seventy-four-year-old cancer patient famously attempts to clarify, perhaps for the final time, what he has long meant by the term "survival." Survival, Derrida explains, understood in its deconstructive sense, is something structural. It must not, therefore, be confounded with anything simply subjective and it must not, above all, be mistaken for any merely self-interested striving for immortality:

> At the moment I leave "my" book (to be published)—after all, no one forces me to do it—I become, appearing-disappearing, like that uneducable specter who will have never learned how to live. The trace I leave signifies to me at once my death, either to come or already come upon me, and the hope that this trace survives me. This is not a striving for immortality; it's something structural. I leave a piece of paper behind, I go away, I die: it is impossible to escape this structure, it is the unchanging form of my life. Each time I let something go, each time some trace leaves me, "proceeds" from me, unable to be reappropriated, I live my death in writing.[1]

The trace, the trace that "I" leave, this trace that signifies to me at once my death and the hope of this trace surviving that death, this simultaneous and contradictory signification that we call survival, Derrida here states, is something structural. It is bound to the structure of the trace, the finite nature of which necessitates at one and the same time the constant threat of death and hope of life to come. The finite trace that I leave, this trace that, in point of fact, constitutes the "I" and makes possible all acts of self-identification, is itself already structurally a form of survival, and therefore nothing to be confused with the more common, subjectively oriented, and undoubtedly metaphysical ambition of immortality.

On the one hand, Derrida clearly indicates, there would be the trace and the structure of survival that comes along in its train, with its constant coimplication of life and death arising out of the play of differences without whole parts. On the other hand, there would be what he here names "a striving for immortality," which should under no circumstances be confused with the former, insofar as it would seem to involve a total effacement of death, and thus of life as well. The thinking of survival is therefore not a matter of immortal ambition. It is not an ambition of immortality, this lust or desire for a deathless state that has so marked the Western philosophical and literary traditions, and done so at least as early as Plato and Aristotle, and at least as late as Descartes, Kant, and Hegel, to say nothing of the variations introduced by Christianity, psychoanalysis, the life sciences, cybernetics, and, most recently, transhumanism.

Such, then, would be the last word of deconstruction on the matter of immortality—if not also, in a rather poetic twist, deconstruction's last word. The ambition of immortality, precisely because it adheres to a logic of the subject—and one, moreover, that would seem to preclude any chance for the event—must be separated at all costs from what has heretofore been said concerning the structure of survival, this "weave of survival" *(tissage de survie)* that—we learn in the second volume of the posthumous *Beast and the Sovereign*—precedes the difference between life and death, precisely so as to open it.[2] There is "survival," a necessarily structural survival that is operative here and now, in or at the very origin of the articulation of these words *on* survival, and then there is "striving for immortality," evidently absent from this scene, or at least from the mise-en-scène of this scene, for, of course, its purely subjective specter must now be conjured to the center stage so as to cast it aside all the better. Deconstruction, it would be tempting to say, is nothing but, and has never been anything but, a deconstruction of immortality, as well as a deconstruction of the lust or desire for this absolute presence that immortality evidently names, and which may, from time to time, appear under the quasisynonymous figures of sovereignty, the death penalty, autoaffection, infinite absolute presence, and God.[3]

With that would appear to conclude, in univocal and monolithic fashion, the position of deconstruction, in the face of what is called (in the various languages indebted to Latin, anyway) immortality. But has deconstruction really no more to say about this matter? And is there really no more to this matter—no more that needs to be said about it, today, in its manifold transhuman, biotechnological, and religious manifestations, than its summary rejection and abandonment, as the other of survival, the false idol of outmoded theology, the narcissistic delusion of technocapital, or the flawed, self-contradictory desire of desire? Is there no afterlife of immortality, in other words, in the wake of deconstruction?

Long rejected on the grounds that it embodies all that is wrong with Western metaphysics—both in its fantasies of political theology and its apparitions of spiritual resurrection—immortality, so the story goes, would be guilty of both denying what is most living in life and contradicting the basic structure of finitude. Immortality—understood variously as the impossibility of dying, the immutable and timeless nature of divine being, the ultimate expression of life's presence to itself, or the indefinite perpetuation of a temporally imperishable biological form—could therefore be nothing but death itself (Blanchot), or, perhaps, at best a term adequate to describe those forms of life incapable of being-toward-their-own-deaths, and thus incapable of the experience of dying, phenomenologically speaking (Heidegger).[4] The desire for, ambition of, or striving after immortality, it therefore follows, must either be considered a misunderstanding (Hägglund)—for how could one desire for the end of desire?—or, at best, the collective expression of a yearning for the end of difference itself (Baudrillard).[5] Any remaining recourse to this name, we might then conclude, must be mistaken, merely metaphorical, or flat-out misguided. To speak of immortality today is, in sum, to do so in ignorance both of its self-defeating nature and of the impropriety of the desire thereby expressed.

The consequences of this line of thinking, which marks the widest range of work not only in deconstruction, critical theory, and phenomenology but also in critical posthuman studies, feminist thought, and biomedical ethics, are multiple.[6] If immortality is simply another name for absolute presence, and if the longing for it is therefore nothing but an ill-conceived desire for the presence of the present, then one need not probe the nature either of the desire, or of the thing, any further. One need not ask, for example, whether such a state, and the desire (or striving) for such a state, constitute two separate phenomena. One need not ask whether there truly would be one immortality, and one could, without further hesitation, dismiss as metaphysical any discourse that remained invested in it, or disregard as deluded any discussion that remained marked by the desire for it. Immortality—in an ironic, though not necessarily surprising, reversal—would therefore turn out to be nothing but a dead end.

It is in the face of such calls, at last to embrace our mortality and give up the dream of eternal life, that *Prosthetic Immortalities* asks whether, in fact, the book on immortality has been closed too soon—whether the foreclosure of the question, be it in the name of finitude, biological structure, or indefinite life, does not precisely betray a repression of immortality's complex, a forgetting of the rich history of its forms, and a denial of the insidious nature of its far-reaching lure. Though Derrida himself at times dismissed the question, in what follows, I show that strictly speaking, deconstruction can be said to be no more against immortality than it is against—or for—mortality, stricto sensu. Far from leading to immortality's pure and simple rejection, the structure of survival entails immortality's renewal and rebirth, and it does so, I show, as prosthesis.

The possibility of such rebirth, in the name of this seemingly timeless lure of timelessness, is therefore no less a consequence of the logic of survival than is its opposite: *la mort sans phrase*. This is why, if we look more broadly at Derrida's writings, we find that some thirty-five years before the final interview, already within *Of Grammatology*, the father of deconstruction was prepared to claim that *différance* does not name a mere "return to finitude": "But it would not mean a single step outside of metaphysics if nothing more than a new motif of 'return to finitude,' of 'God's death,' etc., were the result of this move. It is that conceptuality and that problematics that must be deconstructed. They belong to the onto-theology they fight against. Differance is also something other than finitude."[7] In the wake of the death of God, and, I might add, in the wake of the recognition of immortality's proper impossibility, *différance* nevertheless names something else and something other than an appropriation of being-toward-death. Were it such, or were we to believe that it were such, then it would mean a return to the very ontotheological conceptuality that, precisely, Derrida claims in 1967 is still in need of deconstruction. This is why—to return to 2004—the subject of the final interview must ultimately remain "uneducable" concerning both "knowing-how-to-die" and "knowing-how-to-live," even as he unconditionally affirms life, and with it, survival.[8] To remain uneducable in this way is to acknowledge the persistent enticement of this specter of immortality that, he also knows, he ought to know better than to desire: "It is necessary, beyond the rational animal that we are and that, *qua* animal, does not die, it is necessary for us to become mortal. That's the great lesson to be learned, for the deaf, like me, who keep trying to learn how to become immortal, or a-mortal, basically like beasts. Ah! If only we could stay beasts!"[9]

Starting from Derrida's tête-à-tête with immortality, *Prosthetic Immortalities* explores why deconstruction is, and must remain, at war with itself vis-à-vis the questions of life and death, of living and dying. If *différance* is something other than a return to finitude, then it is because it negotiates with this phantasm of eternal life where other, more philosophically inclined discourses mistakenly believe to be able to be over and done with it. Rereading the question of immortality through the Western philosophical and biological traditions, I show how the Derridean thinking of survival not only calls us to take immortality's promise all the more seriously, now that we believe at last to have surpassed it, but also teaches us to recognize its most far-reaching biotechnological, philosophicopoetic, and zoographic traces, there where we suppose we have left them far behind. Far from dead and buried, the problem of immortality lives on today. It does so by way of physiological and informational accounts of life that carry on a legacy begun at least 2,400 years ago. By entering debates in phenomenology, transhumanism, atheism, and biomedical ethics, *Prosthetic Immortalities* develops a deconstructive theory of immortality—now understood as prosthetic immortalization—which exposes the complicity between the indefinite life of the posthuman and the immortal life of classical ontotheology.

The maturation of this book was aided by many people. Their generosity, be it in reading portions of the book, conversing about its subject matter, or conspiring on a novel cult, was essential in bringing it to its final form. I owe them a great thanks and, more so than thanks, a place in the life of the work. I am most grateful to Nicole Anderson, AJ Baginski, Geoffrey Bennington, Maddalena Cerrato, Patrick Gamez, Deborah Goldgaber, Dinah Hannaford, Steve Hannaford, Jonathan Hay, Thomas Hays, Constantine Karanthasis, Elissa Marder, Michael Marder, Armando Mastrogiovanni, Laliv Melamed, Andrew Mitchell, Alberto Moreiras, Richard Murphy, Bruno Penteado, Michael Portal, Janie and Eric Rosenthal, Kas Saghafi, Mauro Senatore, Deirdre Madeleine Smith, Alexander Thomas, Sergio Villalobos-Ruminott, Teresa Vilarós-Soler, Etienne Vouga, William Verheul, and David Wills for their help. Charlotte Ruth Rosenthal, whose life has largely coincided with the present preoccupation, and who remains my personal best chance for posthumous survival, has been an inspiration throughout. Finally, I'd like to extend a special thanks to Cary Wolfe and Douglas Armato, University of Minnesota Press, for their confidence in this project and their support of it to the very end.

INTRODUCTION

On Prosthetic Immortalization

> There is no body that is not also an articulation, and no articulation that does not imply a radical notion of divisibility. In writing prosthesis, one enunciates first of all such ideas of articulation and divisibility, and, conversely, a drift along the process of adjunction, a prosthetic chain, indiscriminately contagion and promiscuity. That is the shifting ground the operation known as prosthesis stands on.
>
> —David Wills, *Prosthesis*

> No doubt life protects itself by repetition, trace, différance (deferral). But we must be wary of this formulation: there is no life present *at first* which would *then* come to protect, postpone, or reserve itself in différance. The latter constitutes the essence of life. Or rather: as différance is not an essence, as it is not anything, it *is not* life, if Being is determined as *ousia*, presence, essence/existence, substance or subject. Life must be thought of as trace before Being may be determined as presence. This is the only condition on which we can say that life *is* death, that repetition and the beyond of the pleasure principle are native and congenital to that which they transgress.
>
> —Jacques Derrida, *Writing and Difference*

Prosthetic Fictions

At first glance, the expression "prosthetic immortalization" would seem to imply an opposition. To take up prosthetic immortality would be to take up a field, category, or concept opposed to some preprosthetic, pure, or unadulterated immortality. There would be immortality tout court, and then there would be the technoscientific or technographic supplement, affixing to the full presence of the original a newly artificial appendage. In contrast to natural, pure, or unalloyed immortality, one would thus find a modern, denaturalized cyborg counterpart. Indeed, even to the simple, unified, and unique meaning of immortality, one would encounter a multiple, diverse, and ultimately disparate polysemy, the very plurality of which would now make difficult understanding in what the simplicity of the original had initially consisted.

To this first glance, we are by now, following the pathbreaking work of Jacques Derrida, David Wills, Bernard Stiegler, and Donna J. Haraway, accustomed to adding another. The origin, on closer inspection and according to a well-known inversion, turns out to be prosthetic, an "original supplement" that in turn vitiates the very privilege of "the origin."[1] To think prosthetically, to read the vast logic of supplementation, or to trace the play of the trace is thus to come to understand that the very opposition between the before and after of prosthesis is itself deconstructible, a function of the historical sedimentation that one calls metaphysics. Understood in this sense, prosthetic immortality would name the always already fallen, auxiliary, or graphic nature of immortalization. It would name the nonpresence of what had always comported itself as though it were immutable, selfsame, ahistorical, and imperishable, but only by ignoring, repressing, or marginalizing some problematic other.

This book is invested in a fiction, or better yet, a series of fictions: the fiction of immortality as that which, on the one hand, must precede any and all prostheticization, and the fiction of immortality as that which, on the other hand, could only follow from the infinite extension of prosthesis. Above all, however, it is invested in the possibility of the translation of the one into the other, and back again. That is, it is invested in the possibility of passing from the fiction of the pure, deathless life of the gods and immortal souls to that of the grotesque, biotechnologically enhanced, undead, and uploaded brain.

The interest of such a possibility of translation lies in the particular purchase that biotechnological fiction holds today. With the move toward technoscience, immortality appears at last to have turned a new leaf. Finally disburdened of religion, superstition, and ontotheology, the dream of immortal life can now be established on rational grounds. Its realization can in turn be engineered, its possession secured through ready money. The names under which immortality now circulates include the Singularity and Superintelligence, the 2045 Initiative and the clone, the posthuman, transhuman, and neohuman, and the engineered elimination of the biological necessity of death.[2] Thus, both the meaning of immortality and the paths by which one hopes to achieve it have today been radically reenvisioned. Such fundamental transformations owe their possibility to developments in information theory, molecular biology, computational neuroscience, and hypercapitalism. With modern modes of experimentation at its disposal and empirical models available to serve as its basis, immortality seems finally to have risen from the ashes, fully reborn.[3]

It is the task of what follows to analyze the structure of this turn. In doing so, we will attempt to reconstitute the link that runs between each of these discourses and their respective fictions. We will show how, in spite of every novelty of invention, the fundamental principles of projects of radical life extension—dedicated to the extreme lengthening of human life, be it through cryogenics, full brain emulation, biological engineering, bespoke and prosthetic organ transplantation, or antiaging technologies—remain bound both to the oldest of metaphysical temptations and to the very same aporias already confronted by Plato and Descartes. To read immortality prosthetically is to render intelligible this link.

The starting point for such a project begins in a commonplace so obvious as to remain virtually unquestioned by the Western philosophical tradition, to say nothing of its scientific and literary counterparts. If there is one thing we know—if there is one thing that most everyone, including Plato and Aristotle, Descartes and Kant, Hegel and Heidegger, and even Bostrom, Harari, and Hägglund, can agree on—it is that we—we who read and write, dream and think, act and love—are mortal. It follows that immortality, or its promise, must then lie on the other side of biologically, or temporally bound, finite consciousness, ontologically, historically, and technologically set apart. Immortality, like the grace that Kleist's dancer extols, would therefore only return on the other side of infinity:

> "Grace appears purest in that human form which has either no consciousness or an infinite one, that is, in a puppet or in a god."

> "Therefore," I said, somewhat bewildered, "we would have to eat again from the Tree of knowledge in order to return to the state of innocence?"
>
> "Quite right," he answered. "And that's the last chapter in the history of the world."[4]

For Kleist's beloved Mr. C, the attainment of grace, be it in absolute self-presence or absolute non-self-presence, is always understood to require the supplemental work of some prosthesis. Neither puppet nor god, the human must become one or the other, lest it remain forever graceless, which is to say forever limited by its lack of absolute limitation (puppet) or absolute illimitation (god). The human stands between these poles, at an infinite remove, ontologically speaking, from both, and therefore forever awaiting a supplement capable of realizing either apotheosis or reification.

What goes for Kleist, writing at the beginning of the nineteenth century, goes for a large part of Western tradition. If one thing is apparent, it is that we are—and indeed must be—mortal. Salvation, revelation, immortality, and even indefinite life must lie on either side of our current state of finitude because it is only for the infinitely simple or infinitely complex that anything worthy of these names might be possible. From the Platonic doctrine of the immortality of the soul to transhumanist dreams of consciousness uploading, cryogenics, desenescence, and cloned offspring, *Prosthetic Immortalities* traces the promise of the fiction of immortality from both sides of the tree of knowledge. It argues that even if immortality is, in its classical Platonic, Christian, or Cartesian configurations, what ought to precede or exceed difference and differentiation, strictly speaking, there has never been immortality without prosthesis, because there has never been immortality without trace. One must write, inscribe, or otherwise perform the fictional production of an other time, place, or mode of being, where immortality's true referent is always, in the final analysis, waiting to be found. Immortality, it seems, has never presented itself in the flesh—that is, outside the oracular words of priest and prophet, or the inspired visions of poet and philosopher.

To read the history of immortality is for this reason to read the history of its prostheticization, be it in the mode of systematic effacement (theology) or, most recently, technological overbidding (transhumanism). In spite, then, of great differences in method, theoretical foundations, and technoscientific conceptions, the necessity, in each case, of positing a prosthetic supplement will be seen to link these efforts. What must be asked, therefore, in addressing this history is how and why the nature of the supplement itself has been transformed.

Phantasms of Survival

Prosthetic Immortalities reinterprets Western discourses of immortal life through the notion of prosthesis, as it has been developed, above all, in the work of Jacques Derrida and David Wills. It reads iterations of immortality as written, which is to say engaged in processes of translation, supplementation, and symbolic forms of compensation. It then asks, on that basis, how these figures have played (and continue to play) decisive roles in the formation of our most basic and fundamental notions of self, identity, individuality, species, and lineage. If there has never been immortality without trace, then there has never been a simply immortal immortality. This difference at the origin of immortalist discourse in turn allows us to read between classical and modern accounts, as well as theological and secular ones, and to ask in what the bond between belief in the immutable being of a soul and hope for a future, data-driven existence may consist.

Yet demonstrating the essentially prosthetic nature of immortalist discourse is only the first step of our project. The second consists in a closer examination of mortality, of the link between its various ontological and biological conceptions, and of the corresponding visions of immortality that they allow to flourish. Thus, while Part I shows how, already in Plato, immortality is understood to be the prosthetic production of a life that cannot but negate its difference, Parts II and III explore how conceptions of mortality and immortality are coconstitutive. By focusing on Heideggerian ontology in Part II, and next on biological approaches to life in Part III, we will be able to address two questions that dominate talk of immortality today. First, what is the relation between religious visions of immortality, above all of God and the soul, and modern biological conceptions of life, the latter of which evidently ground, for the first time, the empirical possibility of something like indefinite life, understood as a physiologically based existence with unlimited duration? Second, if mortality constitutes the essence of our being, as today seems undeniable, then whence the desire for immortality? Far from a simple error of conception or a mere failure of fortitude, the prosthetic birth of immortality is written into the fabric of life and existence. The upshot of this is that no amount of naysaying will be sufficient either to eliminate its lure or to stop it from taking on new and unheralded forms. I explore such forms, as well as the difficulty of simply dismissing them, in Part IV, when I engage directly with the ideas of antiaging, whole brain emulation, and cloning. Prosthesis, in sum, constitutes the law of immortality, which only ever is as the iteration of its fictions. Yet recognizing that there are only ever fictions of immortality is, in itself, a meaningless gesture, lest one also link the production of these fictions back to the most basic and fundamental processes constitutive of (mortal) life. Reconstituting this link by bringing together phenomenological and biological insight is the overarching goal of the four parts of this book.

The concept of prosthetic immortalization developed in Part I, through readings in Plato and Derrida, provides a means of passing between current trends in posthumanism and transhumanism—cryogenics and biological engineering, cloning and brain uploading—and canonical, ontotheological conceptions of eternal life—immortality of the soul, of fame, of the gods, of God. This passage is possible because, as I argue in chapter 1, even the most dogmatic belief in the soul's immortality must at some point make recourse to the inscription of this belief, and thus to a logic of the trace or supplement, even if a repressed one. In this way, by way of this shared recourse to technics, not only may biotechnological trends be illuminated by classical, ontotheological thought but also these trends may themselves illuminate unrealized aspects of this classical tradition, hitherto unremarked. Only when this divide has been bridged and its deep continuities exposed will the properly historical dimension of contemporary efforts at radical life extension in Itskov, de Grey, More, Minsky, Chalmers, and Kurzweil become visible. Only then will their true novelty become apparent and the performative (or effective) nature of all immortalist discourse be revealed. Far from a ground for dismissal, the link that joins these traditions is the key to understanding their true gravity.[5]

Immortality has thus always been prosthetic. Yet were it simply a matter of these instituted, immortalizing traditions—one evidently religious and the other secular, with one locating immortality in a prehistoric divinity and the other in a posthistoric technoutopia—then one might well have expected to have done away with the problem by now. Were immortality simply a matter of reason or instituted faith, then one could have presumed to have dispelled its fictions and eliminated the concomitant confusions. This is precisely what philosophers such as Martin Hägglund and Nick Bostrom have attempted, in the former case by embracing a discourse of death, in this life, and in the latter case by asserting an apparently more rational alternative to classical conceptions of immortality, through what is today referred to as indefinite or amortal life. In other words, indefinite life, although not properly immortal—because still susceptible to death—is nevertheless no longer biologically fated to die.[6]

The secular efforts of Hägglund and Bostrom thus stake their claims against what they understand to be the faults of generations of ontotheological thought. They do so, moreover, as though from positions outside the purview of those errors. Yet for this very reason, their work raises other, no less problematic questions. In Hägglund's case, it bears asking whether one can in fact simply embrace death. In Bostrom's case, one must wonder whether evading the pitfall of immortal life is as straightforward as opting instead for the most (mortal) life possible. Though the positions of Hägglund, proponent of radical atheism, and Bostrom, founding father of contemporary transhumanism, are antithetical on the surface, both the mortalism of the former and the utopianism of the latter depend on the same presupposition: a clear, simple divide between the desire for mortal life and the desire for immortal life. Each wishes to be free from the lure of absolute presence—to be free, in sum, of religion, and to be free of the ontotheological illusion of life immutable and eternal. It is by way of such freedom that each claims to reject immortality—along with its narcissisms, its confusions, and its obfuscations—so as to embrace more rational, more finite, more ethical, or more human ends, made possible by the cogent knowledge of finitude. Both Hägglund and Bostrom know that immortality is impossible. Yet despite the fact that, to all apparent evidence, one has always known this, both the word and the lure remain. By what necessity?

Immortality, though impossible, unthinkable, and even undesirable, remains an active and unavoidable feature of the discourse of mortality, even where one recognizes that it does not belong. Where contemporary critiques of immortality ultimately err, be they in the name of finitude or indefinite life, is in failing to recognize that their own insights have always been available—that, in fact, both the word "immortality" and the thing to which it refers have always been out of place, that one has always recognized this about it, and that therefore its sense has never been as univocal as one today likes to think.

The attribute "immortal," and with it those of αθάνατος, *immortalis,* and *unsterblich,* have always been in the process of displacing themselves. They have always been extended, far beyond their allegedly proper sense of immutable, deathless existence, into evidently secondary and merely figural uses of these terms. This was the case already for Achilles's fame (κλέος), which takes the immortality of the name as a placeholder for that of the man, and it continued to be so for the many poematic afterlives subsequently promised by Horace, Shakespeare, Wordsworth, and many others. These writers were no more deluded as to the immortality promised by their poematic supplements than gerontologist and life-extension advocate Aubrey de Grey is about the current prospects for ending aging—which is not to say that both are simply deluded. Rather, it is to recognize that visions of immortality, whether strictly realizable or not, and whatever their basis, be it religious, poetic, or empirical, are never simply measurable by objective standards of truth and falsity. This is because these efforts do not appeal to ready-made notions of life and death (against which they could be measured) but are instead involved in reshaping these very notions. Immortalizing efforts thus cannot but inform, in an essential fashion, the very object that they claim merely to extend or preserve.

Homer's promise of immortality through fame thus already constituted a complex discursive operation. It was both a claim on eternity and a recognition of the limited nature of that claim, as well as a modification of the relation between the two. Such complexity remains no less at work today. It is present in the declarations of the biological immortalities of bacteria and stem-cell lines, as well as in the informational immortalities of genes and still wholly theoretical, inorganic brains. In each case, one promises immortality, only then either to acknowledge the failure of that promise as soon as it is uttered or to displace the sense of immortality to which one had appealed. What must be addressed, therefore, is the meaning of the possibility of such displacements. What does it mean for Western ontotheology and technoscience that the very term to which metaphysics seems oriented—immortality—neither is nor has ever been stable? Is this instability congenital or contingent? Can one prevent the (inevitable) movement outside the proper of immortality (toward the prosthetic), or, conversely, back inside, toward the (preprosthetic) lure of atomic indivisibility? It is telling that in the case both of poets and transhumanists, specters of immortality always seem to straddle the ontotheological and the technoscientific through promises of translative, textual afterlives. In both cases, it is by way of some code that immortality should become attainable. It is through the fiction promised and performed by some code—be it linguistic, genetic, or cybernetic—that some life will be able to survive.

In what follows, I argue that the sense of immortality has always been in flux. This flux has perhaps received more publicity recently, as developments in bioengineering have seemingly upended our security in the necessary limits imposed by biological corporeality. But for all that, it is not something new. What must be tracked is the way that this flux is now mobilized, crossing discursive borders and proliferating novel fictions. Today, the plasticity of the promise of immortality traverses both the fictional accounts of recognized fiction authors such as Houellebecq, Shteyngart, DeLillo, and Ishiguro as they explore the dystopian futures of an ever more malleable posthumanity and the (no less) fictional accounts of recognized academics and entrepreneurs, such as Moravec, Minsky, Bostrom, Sandberg, Itskov, de Grey, Istvan, Thiel, and Page, as they speculate on the ever-proliferating new toys of technoscience. It has made thinkable the production of novel genetic fictions, in the bacterial chimeras of Christian Bök, the so-called deextinction of wooly mammoths, and the potential immortality of self-rejuvenating jellyfish, *Turritopsis dohrnii.*[7] Alongside such inventions, semantic transformations are also taking place, ones internal to the very concepts of immortality, life, death, and mortality, as developments in genomics, informatics, cloning, bespoke organs, homeostatic soft machines, artificial respirators, and synthetic brains generate new discursive and prosthetic possibilities.[8] These events are of great significance. Nevertheless, such modern marvels should not obscure the more fundamental point that the slightest historical attention may bring forward—namely that the question of immortality has always been an open one, and immortality itself always already indefinite. Immortality has always informed forms of survival, and vice versa.

Such a possibility of transference and translation, built on the porous border between (indefinite) survival and (infinite) immortality, and grounded in the textual structure of life itself, is the premise on which both the Western philosophical and poetic traditions take shape, in Plato and Homer, respectively. Without such a foundation, neither Bök's biomolecular manipulations nor Bostrom's and Chalmers's informational clones would be intelligible.

Thus, however paradoxically, what the demonstration of the logic of prosthetic immortalization at work within the ontotheological and technoscientific traditions in fact reveals is the persistent imposition of eternal life's most menacing and metaphysical lure, even there where one knows—or ought to know—better. The desire for a selfsame, death-defying ipseity persists in spite of, and indeed due to, the very conditions that ought to make it impossible in the first place. This is why the persistence of this desire, once reduced to its most essential elements, cannot be isolated to the baldest projections of atemporal, eternal being, be they religious or secular, modern or ancient. Where proponents of atheism, philosophical transhumanism, and phenomenological finitude read the traces of immortalist discourse as mystified, cynical, or merely mistaken, believing that they can purge themselves of the errors of those deluded multitudes who desire both what is impossible and what is undesirable, they have not only failed to read the tradition but also to account for the structure of their own discourses.

One of the central tasks of *Prosthetic Immortalities* will thus be to reappraise what is meant by the desire for immortality in order then to situate its basis on more fundamental grounds. Not strictly psychological, religious, or even human in nature, what is called immortality must be located at a more elemental level. Immortality—or, to be precise, the phantasm of immortality—is a fundamental feature of the structure of existence, which is also to say that it is a fundamental feature of the structure of life. Only once immortality has been grounded in this way will it be possible to explain not only its staying power but also the gravity of its always innovative iterations. Far from a merely human or strictly anthropomorphic lure, the specter of immortality must be shown to follow from the structure of life itself. Only in this way can an account be given that no longer naively opposes empirical, biological evidence with phenomenological experience.

Immortality, I thus demonstrate, is a phantasm, and one in the

Derridean sense of the term. This means, first, that although it is itself the effect of artifactual, prosthetic performances, what it performs is just its indivisibility; and second, that it is performed is no mere contingency but a necessary consequence of the structure of the trace.[9] As such, immortality's lure is not susceptible to mere demystification. It is instead an irreducible element both of the discourse on and of the very structure of life, constitutionally inscribed into even the most finite finitude, and locatable both at the scale of biosemiotic recursivity and that of phenomenological being. Insofar as the question of life can be circumscribed by that of world (construed, in the broadest possible sense of the term, as autoaffection or homeostatic reflection), immortality remains one of its irreducible phantasms. Rather than rejecting it, therefore, be it in the name of science or atheism, one must instead learn to negotiate with it. Far from allowing us to eliminate the problem of immortality—the desire for which ought to be compared with its contemporary correlate, the immortalist's desire to solve the problem of death—the acknowledgment of the fundamental and constitutive impossibility of the classical concept of immortality, on the grounds of its artificial, posited, or prosthetic nature, is only the first step in taking its imposition seriously.[10] To read its traces as prosthetic is thus to begin to identify the structure of its ongoing survival. Indeed, and more to the point, it is to mark out what makes its lure unkillable, like everything that is indefatigable, beginning with God.[11]

The problem of immortality will not cease to haunt us. That is the broadest conclusion of this book, and one that I draw out in readings that extend from classical metaphysical to poststructuralist accounts of divinity in Plato, Descartes, and Derrida, but also empirical-biological descriptions of life in Jonas, Jacob, Varela, and Maturana. It will not cease to do so because, as both Freud and Heidegger show, each in his own fashion, far from a mere error, the promise of immortality is itself grounded in the structure of existence. Mortal life, be it through the *Unbewußtsein* or *durchschnittliche Alltäglichkeit,* produces immortality as its phantasm.[12] Thus, both the phenomenological tradition (broadly construed) and biological accounts of autopoiesis ultimately offer a similar lesson: that life is as a process of prosthetic immortalization.

Once seen from this perspective, the question becomes whether one could ever hope, finally, to learn to live without it. If not—if its lure is a necessary structural feature of what ought to preclude it (mortality)—then immortality would call for another mode of analysis, one able to read its traces as prosthetic (i.e., effects produced in and through the structure of existence) while accounting for what of its phantasm is destined to remain active and at work: an unkillable specter, impervious to critical modes of analysis or disillusionment, and functioning especially where one believes to have eliminated or superseded it.

In Part I, I develop a thinking of immortality's phantasmatic nature through readings in Plato and Derrida. While Plato establishes in what consists the most extreme problem of immortality—as guarantor of the unity of life, which is otherwise threatened by pure dissemination—Derrida provides the language and conceptual framework necessary to grasp immortality's ongoing survival. It is no longer a mere vestige of theology but instead an ineliminable element of the structure of life, or what he calls the structure of the trace. To be sure, Derrida would never embrace immortality. In contrast to Heidegger, however, for whom a free, undistorted relation to death remained Dasein's only authentic mode of existence—if not the very privilege of Dasein as the ground of its ontological specificity—for Derrida, Dasein's possibility of attaining to the experience of "its most proper impossibility" is, for essential reasons, held ever in suspense. The possibility of appropriating finitude is therefore, by virtue of the very structure of finitude, rendered impossible—a thought captured in Derrida's infamous contention that infinite *différance* is finite.[13] The play of the trace, we could say, allows neither for (authentic) mortality nor for (authentic) immortality, as such. Once conceived as impossible, the promise of mortality, understood as the appropriation of one's proper impossibility in death, becomes a lure just as great as its opposite. Indeed, it may come to serve precisely as a subterfuge for the latter.

The principal mark of distinction of Dasein therefore ceases to be one in Derrida's reading of the existential analytic, which he develops variously in *Aporias, Demeure: Fiction and Testimony, The Beast and the Sovereign, The Death Penalty,* and *The Animal That Therefore I Am.* Anyway it ceases to be as clear-cut as such a conception—and the hierarchical distinctions that ought to follow from it—would seem to require. This is not to say that, failing to attain to an understanding of death as death, Dasein somehow simply becomes immortal—be immortality conceived ontologically or ontically, as an obliviousness to death or as an infinite extension of deathless being in or outside vulgar time. What is revealed is rather that mortality, understood as a condition or way of being proper to the being that has an understanding of its own being, is for Derrida a metaphysical lure. The self-presence required for the appropriation of the *sein-zum-Tode* becomes just as problematic as the lack of self-presence required for the absolute negation of the *sein-zum-Tode,* in the *Benommenheit* of the animal (or puppet), or the *Unsterblichkeit* of the immortals (or God).

The figure that Derrida chooses to express such a structure of existence that would not succumb to temptations like these is survival. Survival, *survivre, survivance,* and *survie* become, especially in the final years of his life but already as early as 1976, the privileged terms through which a notion of life–death as textuality may be articulated. Survival, as Derrida states explicitly in the final interview, is neither immortality nor a desire for it; instead, it is the originary weave or *tissage de survie* that sets life and death in motion in the first place. Nevertheless, as I argue in chapter 2, if survival makes possible the very opposition between life and death, as the trace does that of presence and absence, then it is not without some relation also to those phantasms of immortality and mortality, which are themselves extensions of this opposition. As Derrida himself repeats, it is the absence of any origin or originary presence that births the desire for origin. This structural absence originates the metaphysical lure, which is why said lure is, in his own words, "ineffaceable," "ineluctable," "indestructible," and "unkillable." It is an effect of a system rather than an arbitrary expression of will, whim, or even phylogenetic history. Just, then, as arche-writing sets into play the notion of and desire for a speech that would efface it, survivance, I show, grounds the very concept of immortality that would deny its bearing, insofar as such a repression of death is built into the self-identity, or ipseity, of the living, from the first trace of homeostatic self-awareness or the inaugural response of a recursive feedback loop. Survival is not simply agonistic—nor even agnostic—with respect to immortality, but gives it life.

The question that follows in the wake of this insight is how best to negotiate with this phantasm. If one will never be able to cleanse survival fully of the dream of immortal life, because the former also comprises that to which it is evidently opposed, then what can deconstructive reading, philosophy, or criticism in general hope to accomplish? If one cannot put an end to the dream of endless, immortal existence, because the fantasy of such an end would itself already betray the finite/infinite opposition that it had hoped to subvert, then what type of intervention might yet be hoped for?[14]

It is these questions that *Prosthetic Immortalities* tries to address by turning in its subsequent parts to Heideggerian phenomenology, contemporary biological discourse, and transhumanist visions of posthumanity. The point of each examination is to show how the problem of immortality is not one that we can simply evade, above all by turning to strictly empirical evidence. Life, or survival, cannot but continue to produce the very ontotheological lures that undermine it, be they in the mode of an everlasting soul, an endless, data-driven, disembodied existence, or a selfsame ipseity that might live on indefinitely, but that might not. In this way, life and death, and by extension mortality and immortality, become prosthetic expressions of a survival that produces them as its play. This is why survival, in the final analysis, is nothing but prosthetic immortalization: the phantasmatic production of a becoming-mortal and a becoming-immortal alike. Survival entails the impossibility of ever arriving at either of these ends, but so too does it entail the impossibility of eliminating them as lures.

Prosthetic immortalization thus names neither immortality nor mortality tout court, but rather the perpetual possibility of lapsing into both from the differing/deferring play of *différance.* But just as what is called death is not exhaustive of "end," and just as figurations of dying are only ever partial expressions of the state of mortality, neither is what we call immortality exhaustive of the structural lure of life's dream of immutable, autoaffective presence to itself. For this reason, processes of prosthetic immortalization are not simply to be found where one projects life as undying or undead. Rather, they occur long before it becomes an explicit matter of life after death; they are already visible in the autopositing of the ipse, just as soon as this prosthetic projection becomes articulated as a site of self-reference and thus a protoorganon of an (always possibly infinitely extendable) presence in or outside space and time. One may thus already read it at work within the self's production and extension of itself, in the infinitely iterable grammatical projection of the impersonal "I," or in the recursive feedback loops constitutive of cellular self-regulation. As a prosthetic point of reference, the nature of which is to be reproducible and thus indefinitely extendable, the simulacral production of the first-person pronoun or the cognition of the single cell is but the first step on the path toward the immortality of the soul. It invents this path no less than does the latter in a (phantasmatic) performance that serves to stabilize its own internal possibility.

What I am calling prosthetic immortalization is therefore also finitude. Its play constitutes no less the impossibility than the possibility of life. If, however, this is the case, then a critical, methodological question is raised. Why continue to speak of immortality or immortalization when I have just avowed their statuses as lures—when, indeed, and as Derrida himself states explicitly in the final interview, the very term that was chosen for the role of rethinking the relation between life and death, "survival," is neither immortality nor a desire for this? Why hold onto this old, deceptive term when other, better ones have presented themselves?

As should be clear by now, prosthetic immortality might best be understood as a translation of survival. It *says* survival, in its Derridean configuration, yet it does so while retaining the very metaphysical term that our judgment tells us we would be better off abandoning. By retaining the figures immortality and immortalization paleonymically, I have therefore made the wager that something more can be gained by allowing them to circulate openly, and that by sanitizing our language we miss, precisely, what of immortality remains. As Richard Rorty points out concerning the difference between simile and metaphor, in terms of veracity, the former is virtually always true and the latter virtually always false. Yet in spite of this, or perhaps because of it, the attraction of metaphor is infinitely greater than that of simile. It will be seen that the figure, word, or concept of immortality has rarely, if ever, been used in the full, rigorous sense of the term. It is rarely, if ever, used with what one would be tempted to call literal earnestness. In spite of the term's frequency of use, one is almost always aware of its misapplication, and in virtually every context. *Prosthetic Immortalities* calculates with this error. If survival might be said to be broadly applicable, without malapropism, then prosthetic immortality, by contrast, already implies a contradiction in terms. In what follows, we shall go down the rabbit hole of this error, which every resolution appears helpless to eliminate.[15] Only in this way can we bridge, or witness what remains unbridgeable, between, on the one hand, the Greco-Christian discourse of immortality and biotechnological fantasies of contemporary transhumanism, and, on the other hand, between each of these discourses and that of contemporary biology, which stands midway between them.[16]

Structure of Chapters

The chapters that follow are divided into four parts. Part I shows how for both Plato and Derrida immortalization is understood to be an ineliminable element of the structure of life. Broadly speaking, immortalization can be understood as the process of translating mortal life into immortal life, and vice versa. Though this act of translation ought to be impossible—the bridging of an unbridgeable abyss—the whole historical problem of immortality might be understood as being caught up in the endless variations of translative efforts that emerge in the wake of this acknowledged impossibility. Translation must thus be moved to the fore in any discussion of immortality. Translation is the most general name for the prosthesis by which life invents itself as the mediation between the finite and infinite. It will thus be seen that although the act of translation is never complete, it never ceases positing its completion—a conclusion that will be drawn out further in the readings of Part IV.

Part II explores the relation between the concept of immortality fleshed out in Part I—as indivisibility—and the various notions of immortality put forward by Cartesianism, Heideggerian phenomenology, and modern biology. By way of this analysis, Part II slowly approaches a critical set of questions. In what consists the modern relation that holds among immortality, mortality, and indefinite life? Can indefinite life, in its biological configuration, be severed from immortality? And what comprises the difference between modern and ancient approaches to the latter question?

As we shall see, the story of the Enlightenment—or, more broadly, of the rejection of religion in the name of reason—is also the history of the death of immortality and the birth of indefinite life. This is the story that Part II helps to tell in its recounting of the transition from Cartesian dualism to physiological and phenomenological approaches to life. According to this understanding, immortality (as indivisibility) would be opposed to indefinite life (as divisibility), as faith, dogma, and mere atheism oppose reason, falsifiability, and radical atheism, and as belief in the immortality of the soul gives over to knowledge of biochemical mechanism. Part II traces the development of this story, exploring the relations that hold between the various approaches to immortality and mortality, and establishing the specificity of each term.

At this point, it will be helpful to clarify the various ways in which the phrase "indefinite life" will be used. Throughout the book, I use the figure in two related, but not identical, ways. On the one hand, as the titles to Parts II and III imply, indefinite life refers to what in the life sciences is known as biological immortality, by which is understood a physiological life-span with potentially unlimited duration. In this usage, which is in line with its sense in transhumanist literature, "indefinite" opposes both "finite" and "infinite" in its marking of a distinct form of temporality. On the other hand, I have at times found it necessary to appeal to the term to emphasize the uncertainty of life itself, the indefiniteness of its being, just as soon as it is subjected to prosthetic extension (which is to say always). This parasitic use of the term is intended to draw out the essential but too often neglected consequences of the admission of the possibility of supplementing, or prostheticizing, life. Once one admits that life can be extended, its referent ceases to be secure. Despite its lofty claims, transhumanism thrives on an implicit restriction of what qualifies as human or posthuman life; this restriction is the task of everything that follows to bring into question.[17]

Part III turns from the transition from theological to secular understandings of life, in and around the Enlightenment, to focus on the forms of mortal and indefinite life that are birthed in the wake of modern molecular approaches to biology. Modern molecular biology does not simply provide an empirical account of life; it also presents an account of the mechanisms by which organisms survive—or perish. In this way, it is thought to provide novel grounds for a specifically biological concept of immortality. It does so, however, only with the help of notions of self and individuality, now grounded in the emergent properties of organisms.

The chapters of Part III explore these mechanisms at length in order to show, on the one hand, the necessity in each case of privileging a certain level of organization above so many others, and, on the other hand, how this privilege—or decision—for certain forms of life constitutes at one and the same time the possibility and impossibility of indefinite survival. The afterlife of immortality lives on this decision, which cannot be made on strictly objective or purely empirical grounds. The participation of the biological sciences in the production of fictions of life and death marks, however, their point of greatest gravity; indeed, as we shall see in Part IV, it in turn informs the more speculative approaches of transhumanism.

In Part IV, then, I show how the discovery of these biological structures leads to the development of further translative efforts at immortalization in transhumanist and posthumanist thought. By reducing life to physiological process, biological accounts of living systems open the door to speculative transhumanist fictions of brain uploading, cryogenics, cloning, and biological engineering, even if most biologists do not actually endorse these fantasies. Revealing this complicity, as well as what it still owes to antique notions of indivisibility, is the goal of this final portion of the book.

One of the central tasks of *Prosthetic Immortalities* is to account for the general possibility of variations in conceptions of immortality and indefinite life. The very notion of an absolute difference between mortal survival and immortal life is itself the product of the ontotheological tradition, which posits this difference precisely in order to transcend it. To radicalize immortality would therefore mean to put these opposing poles back into relation with one another by reading how they have always been related. Such an effort is necessarily twofold. On the one hand, it consists in examining how fictions of mortality and fictions of immortality are produced coextensively. On the other hand, it consists in questioning the possibility of a hard and fast opposition between the desire for mortal and that for immortal life. In this way, one can close the gap between ancient and modern accounts, and clarify the limits of any critical effort.

PART I

IMMORTALITY AS INDIVISIBILITY

What is immortality, and what is its relation to mortality? On what grounds could we speak either of a continuity or a discontinuity between ancient and modern approaches to these questions? In chapter 1, I return to one of the earliest and most influential attempts of the Western philosophical tradition to situate immortality and clarify its importance for the mortal world: Diotima's speech in Plato's *Symposium*. I show how what is there called "immortality" (ἀθανασία) is, from the very beginning, intimately tied to the problem of life and to its investment in reproduction. Plato conceives of life as a mode of prosthetic immortalization.

Such a conclusion becomes inevitable as soon as immortality is determined by Diotima as having ontological priority over mortality. As such, it conditions not only love (and with it all future forms of generation) but also life itself, which cannot but attempt to realize immortality even if, in the final analysis, its realization is deemed to be impossible for it.

The enduring relevance of this Platonic vision is then explored in chapter 2, where it is both reversed and generalized by Derrida. Far from being a dogmatic assertion of divine immortality, what Plato offers is an extreme articulation of the threat posed by its loss, as a ground of being. Immortality names for Diotima and Socrates alike the possibility of indivisibility. Once understood in this way, the Platonic text becomes a critical index for all future attempts to conceive of life, be they philosophical, biological, or theological.

The central question raised by the *Symposium* can thus be put as follows. To what extent could any notion of life, once defined through the role of γέννησις, escape this immortalizing logic, which consists in the simultaneous positing of immortality as life's end and of immortalization as the essence of life's activity? For Diotima, love is the driving force behind life, and generation the vehicle through which that love manifests, allowing mortality to come into relation to immortality. Critically, this relation is forged through subterfuge. Mortal modes of creation such as procreation, composition, and institution building—what she calls "generation in the beautiful"—allow for a finite participation in the infinite. They form the bond between mortal and immortal, yet they do so only because they hold a prior relation to the immortality they lack. Throughout her speech, it is precisely modes of indefinite life that are tasked with establishing this relation. Individual animals, evidently fated to die, participate in immortality through prosthetic extensions of their selves that, precisely, may live on, but are not guaranteed to do so. The question that the *Symposium* raises for modern proponents of indefinite (biological) and amortal life is thus whether a mere reversal of ontological priority—from idealism to physicalism, for example—would be sufficient to cleanse compensatory dreams of indefinite extension of the metaphysical specter of infinite, atemporal, immortal being. Is there any concept of life, in other words, that is not simply opposed to death, and thus invested in a project of its own immortalization? The *Symposium* forces us to ask whether life, once conceived as originary investment, contract, or promise, does not already imply a kernel of immortalization—a question that we shall return to in Part III, this time from the point of view of the life sciences.

1

LIFE AS GENERATION IN PLATO

> For this is the most natural of all functions among living creatures . . . : *viz.,* to reproduce one's kind, an animal producing an animal, and a plant a plant, in order that they may have a share in the immortal and the divine in the only way they can; for every creature strives for this, and for the sake of this performs all its natural functions. . . . Since, then, they cannot share in the immortal and divine by continuity of existence, because no perishable thing can remain numerically one and the same, they share in these in the only way they can, some to a greater and some to a lesser extent; what persists is not the individual itself, but something in its image, identical not numerically but specifically.
>
> —Aristotle, *De anima* (415a27–b9)

> Few phenomena in the living world are so immediately evident as the begetting of like by like. A child soon comes to realize that dog is born of dog and corn comes from corn.
>
> —François Jacob, *The Logic of Life*

At the most decisive moment of the *Symposium,* when the secret of love is finally about to be revealed, Diotima, woman, prophet, and teacher, points Socrates to the peculiar nature of generation. Why is it, she asks, that all animals reproduce? And why, in turn, do they care so passionately for their offspring? What is the link that joins each of these experiences, and not only among the highest, but also the lowest, forms of life? "What do you suppose, Socrates, to be the cause of this love and desire?"[1] Unable to answer her, Socrates begs Diotima to explain it to him. "Well then," she replies, "if you believe that love is by nature bent on what we have repeatedly admitted, you may cease to wonder. For here, too, on the same principle as before, the mortal nature ever seeks, as best it can, to be immortal. In one way only can it succeed, and that is by generation; since so it can always leave behind it a new creature in place of the old" (207c–d).[2] Mortal nature (ἡ θνητὴ φύσις) ever seeks, as best it can (κατὰ τὸ δυνατὸν), to be immortal (ἀθάνατος). It cannot but seek this, clarifies Diotima. Yet at the same time, mortal creatures cannot actually become immortal—at least not in the full sense of the term, which is realized by the gods alone. The key to understanding the nature of generation (τῇ γενέσει) is therefore by way of its role within this fundamental wanting (ζητεῖ). Generation, properly considered, would be the mortal realm's solution to the problem of immortality, as it is posed by the divine. Generation is the "one way only" that mortal creatures "can succeed" in attaining this, their true end—not, however, "by keeping it exactly the same for ever, like the divine, but by replacing what goes off or is antiquated with something fresh, in the semblance of the original."[3]

Henceforth, generation can no longer be understood as a primarily physiological phenomenon. The secret to generation lies instead in its ontological function: as the solution of mortal nature to the elusive character of immortal being. That is, generation is the manifestation of an innate, meta-physical yearning, the explanation for which could not be found in corporeal features alone. As such, generation becomes a path shared by all living beings, of male and female gender alike. In one of the most radical passages of the *Symposium,* Diotima had earlier explained to Socrates how "begetting" and "engendering," though apparently the province only of the latter, are in fact affairs common to both man and woman. Privileged forms of divinity within the profane world, begetting and engendering are acts fundamental to all mortal life: "The conjunction of man and woman is a begetting for both. It is a divine affair, this engendering and bringing to birth, an immortal element in the creature that is mortal and it cannot occur in the discordant. The ugly is discordant with whatever is divine, whereas the beautiful is accordant" (206c–d).[4] "Bringing to birth" (ἡ γέννησις), along with "begetting" (τόκος) and "engendering" (ἡ κύησις), thereby come to name for Diotima and Socrates the exemplary forms of manifestation of immortality within mortal life.[5] They are a "divine affair" (τοῦτο θεῖον τὸ πρᾶγμα), and it is only through them that life-forms both great and small, both male and female, may acquire a taste of this "immortal element" that they otherwise lack. Why, we could ask, does the creature that is mortal want immortality, in the first place? What is the principle to which Diotima refers above? Well, the prophet might respond, τό ζῷον wants immortal life because it is only in immortality that it may truly possess the good. Only through immortality may the good come "to be one's own for ever" (207a), and mortal life is, in its Diotimean construal, nothing but such a desire: "From what has been admitted, we needs must yearn for immortality no less than for good, since love loves good to be one's own for ever. And hence it necessarily follows that love is of immortality" (207a).[6]

To be mortal, we might conclude, simply means to strive to attain the good, and not simply to attain it now, but always, ἀεὶ, in the most complete, protracted, absolute fashions that each mortal being can imagine. Mortal life, being incomplete and finite, cannot but desire this plenitude that it is not. The reason for this is that this absence in fact orients the being of τό ζῷον in advance. Life, all mortal life, is driven to supply this lack whose absence conditions it in an originary fashion.[7]

The vision of life that Diotima puts forth in the *Symposium* comprehends the living as an originary prosthesis of immortality. Life—all life—is constituted for her as self-supplementing immortalizing prosthesis. As such, this life is destined to seek an end that is all but impossible for it to attain, and it does so, as we shall see, through forms of indefinite life, which serve it as so many compensatory symbols of plenitude.

In putting forward such an account, Diotima does not simply dogmatically suppose the existence of an immortal, divine realm. On the contrary, she reveals the epistemological and heuristic necessity of presuming such a mode of existence, if "generation" is to be conceived philosophically. The philosophical concept of generation, as a form of reproduction, depends on the prior being of divine immortality if it is to be minimally comprehensible. In this way, her speech brings to light the meaning of immortality—not simply as a form of indivisibility but as the very ground on which indivisibility as such becomes conceivable. For generation to appear as generation, which is to say as a form of reproduction, there must be some immortality. This is Diotima's lesson. Its dividends are vast because it not only establishes the basic model of immortality for the Western tradition that follows but also reveals the inductive process through which this model may come to reconstitute itself each time anew.

Lessons in Love

Diotima's lessons in love are profound and provocative, stunning in their simplicity and captivating in their grandeur. One of the single most influential women in the history of antique philosophy, the priestess helps usher in a new era of Greek thought: transitioning from Socratic dialectic to Platonic idealism, and with it forming the bridge between classical paganism and Christian theology. Her words, which are initially related by Socrates before being retained by Aristodemus and finally recounted by Apollodorus, visibly transform the totality of life, altering both its concept and its scope. In the process, they also irrevocably reconfigure the very project of Western philosophy.[8]

Nevertheless, as no reader of Plato's *Symposium* can fail to recognize, the great irony of her lofty vision lies in the impossibility—or at any rate the probable, all-but-absolute impossibility—of its realization.[9] If becoming immortal is the end to which life is called, then failing to become immortal will be the destiny of all mortal beings who succumb to this, their nature. This is because for the mortal, immortality is the one thing excluded in advance. For the mortal, being alive precludes the realization of immortality, even if it also invents it as an end[10]—whence the beauty and paradox of life as she presents it.

Of course, this is not all that is at stake in her speech. If all that she had asserted was the impossibility of becoming whole and complete, then she would have introduced nothing new within Plato's dialogue that Aristophanes's myth of man's origins had not already put forward. Diotima's innovation—and the particular interest of her vision, both in the *Symposium* and for the tradition at large—lies instead in the logic of compensation that she introduces, and introduces at the decisive moment of the mortal's failure to attain the good, to become whole, and to be immortal. As Diotima has it, in the wake of the constitutive inability of mortal life to attain that which it desires, an entire compensatory system of symbolic participation emerges that allows a trace of this immortality to appear and persist in the world, precisely for those excluded from it. "The mortal nature," she specifies, "ever seeks, as best it can, to be immortal" (207c). The importance of this qualification cannot be overstated. It transforms what might have been viewed as a simply impossible task—and one that could, for this very reason, merely have been spurned—into a project whose impossibility is modified and mollified by an infinity of (entirely possible) consolations. Though it cannot be immortal, the mortal creature can strive for immortality, and it can participate in it through the divine affair that is begetting in the beautiful. To put it better, and more rigorously: mortal life is itself always already this trace of immortality. It is as, and only as, the expression of a constitutively absent immortality whose vacancy sets it into play and preorients it, in advance, as essentially loving (which is to say, engaged in giving birth). Mortal nature participates in immortality "as best it can" or "according to its ability" because the practice of its "ability," though incapable of attaining its ultimate end, remains thoroughly satisfying in its activity.

By inscribing the lure of immortality into the fundamental constitution of the mortal, Diotima thereby invents life as a process of prosthetic immortalization. The aim of immortality does not fall from the sky on the living. Rather, it is inscribed into the being of the being that is mortal, and as an outside that is already inside. At the same time, and despite the fact that she presents mortality as being opposed to immortality, one immediate consequence of her vision is precisely to close the divide between these terms. To the extent that one can compensate for the absence of immortality through forms of what today we might call survival—to that extent, the difference between the finite and infinite ceases to function. Diotima articulates a concept of life that requires immortality in order to be legible. In this chapter and in chapter 2, I will show why this is and how it may persist far beyond the ambit of Platonism.

Recall that, prior to Socrates summoning Diotima to speak before this all-male audience, Aristophanes had told another tale of human nature's origins: that of Zeus's cleaving of the first men. Originally, Aristophanes recounts, humans were twice what they are today, and they comprised three genders: male, female, and male–female. Afraid of their power, Zeus had been forced to divide each member of the race in two. By cutting each in half, Zeus thereby split their strength, effectively eliminating their threat. Now, instead of rebelling against the gods, it was the memory of their loss, inscribed as it was on each of their bodies, that left these half men yearning only for union. Love, as it is commonly called, was born with the birth of this new desire. It named for Aristophanes the desire for the unification of the once-whole body.

But Zeus's plan, we soon learn, had worked too well. For now, and to the god's utter dismay, when these poor dismembered beings actually succeeded in finding their lost halves, they were so satisfied by the reunion that they neglected both food and drink. They simply embraced and would do so to their very deaths. In place of feuding restlessly with the gods, humanity found a bliss too absolute to leave room for much of anything else. It was for this reason, Aristophanes recounts, that Zeus was forced to intervene a second time.

Unsatisfied with humanity's newfound sense of complacency, the god of thunder had no choice but to perform a second act of body modification, reengineering the already-split individuals' physiques to better suit his own ends. This time, with a single stroke, he now eliminated the humans' fresh sense of contentment, in the process producing conditions for a sustainable cycle of life and death. He did this by turning their genitalia around, bringing them from their backs to their fronts, and allowing for the advent of sexual reproduction. Now, upon meeting, any reunited men and women would, through inadvertent but nevertheless efficacious necessity, both produce offspring and experience temporary (but only temporary) satisfaction.[11] What followed was an economy of life, death, and reproduction that functioned to keep humanity both in line and alive, with no concerns past their own insatiable sexual desires and the care of their new, unplanned progeny.

The critical difference between Aristophanes and Diotima's visions thus lay in the form, and indeed the logic, of compensation that follows from each. While Aristophanes's myth of love's origin may have been able to account for basic human reproductivity and sexual preference, it nevertheless created a human destined to remain bound to the specific conditions whereby it was produced. Love, as a mere drive for physical union—indeed, as a mere drive for reunion—remained hamstrung. As Heidegger might put it, it produced a thoroughly *benommen* or *gehemmt* individual, no better really than an animal. Allowing only for the perpetuation of the species through a reproduction destined to reproduce itself, Aristophanes's love left humankind fixed within a closed circuit, without the possibility of history or event. This is surely why Plato found it so unsatisfactory.

By contrast, love for Diotima can no longer be construed as the mere desire to re-member dismembered parts. Because its origins are to be found in an end that already transcends corporeal embodiment—and an end that conditions that embodiment through what is constitutively lacking in it—it is essentially insatiable. Having placed the origin of love beyond the reach of any prior, historical experience of trauma or loss, love may now seed itself within all the variegated, and indeed unlimited, activities with which mortal life finds itself involved. More to the point, it may now even begin to account for those activities in their very disseminal variation.

Love, in Diotima's telling, originates in the incompleteness of the structure that constitutes the living, as mortal, as such.[12] "The mortal nature ever seeks, as best it can, to be immortal. . . . It is only for a while that each live thing can be described as alive and the same" (207c).[13] "So do not wonder if everything naturally values its own offshoot, since all are beset by this eagerness and this love with a view to immortality" (208a–b).[14] "All men are pregnant, Socrates, both in body and in soul: on reaching a certain age our nature yearns to beget" (206c).[15] As Diotima explains, time and again, as though to a child too young to understand, it is less any individual that can be said to be the source of love—love does not come from me, qua experiencing individual—nor even from any distinction of specifically human nature, but from life itself, "the living that lives" (τῶν ζῴων ζῆν), which is *as* beset by love, and is therefore always already such. Insofar as all mortal life by nature procreates, it must also be fundamentally marked by this capacity for pregnancy, and it must be so before anything we might generically recognize as begetting, or even eros, appears as such.[16]

Nevertheless, the originality of Diotima's account is not to be found in this insight alone, which might just as well have led to a purely fatalistic vision of mortal life, given that the one aim of said life is now found to be all but unattainable. Rather, her true innovation lies in the solution to life's impasse that this fatality allows her to discover. The division between mortal and immortal life is both absolute and solute. It is both unbridgeable and susceptible to bridging, for, though I may be thoroughly mortal, I can nevertheless bring a bit of transcendence into the phenomenal world. I can figure said transcendence within the finite limits of my mortal existence, and I can even, depending on who I am, under very special and singular circumstances, and perhaps with a bit of luck, even hope to transcend those limits as such.[17]

It is from the impasse constitutive of mortal life—its inherent, structural desire to become whole, complete, immortal—that is therefore born another mode of compensation altogether, unless this is the very origin of "compensation" as such, understood as the possibility for locating equivalence in difference through the mediation of symbolic structures. What more originary scene of supplemental or virtual replacement, of a pure presence by a mere re-presentation, could there be for living beings than the one built into the very structure of the living qua living? Given the impossibility of becoming immortal, the mortal must, according to Diotima, do the next best thing. Therein lies her singular insight: the notion that there might be a next best thing. Given the difference between mortality and immortality, the fact that one can supplement for this insatiable lack, that there is any form of compensation whatsoever whereby divine immortality might find a path into the mortal world, becomes the decisive step. The living, therefore, being mortal, must settle for partaking (μετέχει [208b]) in the closest corporeal equivalent to immortality. Indeed, it must settle for the very possibility of equivalence and semblance, which serve henceforth as subterfuges (τῇ μηχανῇ [208b]) for the self-present being of the divine (which is to say, for that which is, by definition, without equivalence, appearance, or semblance).[18] It must do so there where the closest mortal equivalent to immortal life may be found. That is, explains Diotima, the nature and role of "engendering and begetting upon the beautiful" (206e). Love in turn comes to name the undeniable driving force behind, and pathway forward for, the mortal on its way to immortality, by way of the subterfuge of generation. Indeed, "love," the very emergence of love, must already be understood to be a form of participation in what is constitutively excluded from the being that loves:[19]

> "For you are wrong, Socrates, in supposing that love is of the beautiful."
>
> "What then is it?"
>
> "It is of engendering and begetting upon the beautiful."
>
> "Be it so," I said.

> "To be sure it is," she went on; "and how of engendering? Because this is something ever-existent and immortal in our mortal life. From what has been admitted, we needs must yearn for immortality no less than for good, since love loves good to be one's own for ever. And hence it necessarily follows that love is of immortality" (206e–207a).[20]

In providing her explanation of love (which, most commentators agree, is also that of Plato, and the moment, verified by the *Phaedo, Phaedrus,* and *Republic,* of Plato's proper break from Socrates with the development of his theory of the forms), Plato, by way of Diotima, by way of Socrates, by way of Aristodemus, by way of Apollodorus, achieves a double victory.[21] The philosopher clarifies not only the nature of eros in the broadest possible terms but also that of generation. What for Aristophanes could only be a secondary effect of a project of divine eugenics—Zeus turning the sexual organs around in order to create a libidinal economy of individual finite pleasures for his human subjects—now becomes written into the very nature of human, animal, and plant kinds, clarified with all the rigor of a fully fleshed-out conceptual constellation. Generation (γένεσις; γέννησις)—both the desire for and the act of, and understood in the full breadth of its diverse semantic applications, from poetic making to biological procreation and from male pregnancy to female engendering—is, simply put, the expression of love.[22] Love in turn is the origin of generation. Yet it is only in light of the immortality that preorients mortal life, only in light of the immortality that at first appears to be the mere condition of true happiness (εὐδαιμονεῖν [205d]) before it is revealed to be its ultimate referent, that it becomes possible to understand how these various and sundry forms of γέννησις are in fact one.[23] United by this single, inescapable movement toward deathlessness, all mortal forms of creation, be they human, animal, or vegetal, thus finally become legible as the continuous expressions of a single, general tendency, toward eternal, selfsame being:

> Well then . . . if you believe that love is by nature bent on what we have repeatedly admitted, you may cease to wonder. For here, too, on the same principle as before, the mortal nature ever seeks, as best it can, to be immortal. In one way only can it

> succeed, and that is by generation; since so it can always leave behind it a new creature in place of the old. It is only for a while that each live thing can be described as alive and the same, as a man is said to be the same person from childhood until he is advanced in years: yet though he is called the same he does not at any time possess the same properties; he is continually becoming a new person, and there are things also which he loses, as appears by his hair, his flesh, his bones, and his blood and body altogether. And observe that not only in his body but in his soul besides we find none of his manners or habits, his opinions, desires, pleasures, pains or fears, ever abiding the same in his particular self; some things grow in him, while others perish. (207c–7e).[24]

> Every mortal thing is preserved in this way; not by keeping it exactly the same for ever, like the divine, but by replacing what goes off or is antiquated with something fresh, in the semblance of the original. Through this device, Socrates, a mortal thing partakes of immortality, both in its body and in all other respects; by no other means can it be done. So do not wonder if everything naturally values its own offshoot; since all are beset by this eagerness and this love with a view to immortality. (208a–b)[25]

If Diotima's gesture at first appears to radically level both sexual difference and ontological difference, extending pregnancy (τόκος), (re)productivity (γέννησις; γενέσει), and love (ἔρως) to all living things, irrespective of sex or species, one nevertheless soon realizes that love is in fact the very basis on which these differences first become marked. What makes man man is not that he participates in love, but rather how he does so, and likewise for woman, lizard, plant, philosopher, poet, and so on. This is why it is so critical to learn about love—to learn, that is, philosophy, which is evidently the only means available to access a pure, unadulterated experience of beauty, one stripped of every material content. What is proper to man is to love beauty itself; it is to beget on the very form of the beautiful in its unalloyed, pure, immutable, and indivisible unity. Hence, it is to ascend from the various lower forms of pregnancy and generation up to the only one that truly befits him, because it is the only one that offers the thing itself in place of an infinity of facsimile. It is the only one, therefore, through which man might hope to in turn engender something truly immortal and divine.[26]

In this way, what Plato offers, by way of Diotima, is a theory of incremental immortalization by prosthetic means. Life cannot but prostheticize itself. It cannot but supplement its originary lack and participate in techne and simulacra. It cannot but engage in (artificial) modes of semiosis and be marked by the life–death that they entail. The mortal, destined as it is to die, can only simulate its immortality. It does so by way of subterfuges: by producing extensions of itself in the world through which it may experience an inkling of this infinite survival outside of it. These prosthetic projections are many, and they are as variegated as are the modes of generation. Critically, they function both for those animals capable of forming a concept of this immortality and for those (we assume) that are incapable—both, that is, for beings that possess knowledge of their desires as such, and those (apparently) ignorant of them. A bird partakes of immortality by producing another bird. A man, by contrast, may opt to father laws or more men. The interest of Diotima's account is that it proposes that both acts not only respond to the same fundamental impulse (love) but also, in doing so, that they function according to the same logic of participation, regardless of each organism's degree of psychic (or cognitive) sophistication.[27]

Yet the above insight contains a further, troubling consequence, one that draws into question both the "naturalness" of reproduction as Aristotle saw it and the "immediate evidence" of begetting as Jacob did, in the epigraphs above. If we follow the logic of participation to its necessary conclusion, then what in fact becomes evident is that each being must have had some access to divine immortality before they could have had any felt experience of it, let alone any explicit desire for it. It is only because both acts of generation (the bird's and man's, but also, it follows, the plant's) have already been touched by immortality that either can come to pass. It is only because living beings have already been preoriented by this drive for immortality, only because the desire for it has been imprinted, in advance, in the being of the living, that anything like generation can arise in turn. This means, however, that counter to our experience—which provides so many examples of generation, through which we first seem to come to recognize the desire for immortality at work—counter to our experience, then, which goes from generation to immortality, it is actually this prior relation to an immortality nowhere present that subtends these modes of generation. It is this prior relation to immortality that makes possible both the emergence and recognition of generation as generation. In other words, without a seed of immortality, there could be no generation—or anyway there could be no generation thinkable as generation. The recognition (evidently so immediate for us) that generation serves to reproduce is in the final analysis only possible because a prior sense of selfsame immortal being informs this act of cognition, turning generation into the figural representation of the immortal's literal presence.

Naturally, generation could perhaps still happen without this seed of immortality, in the sense that animals might still produce offspring, even if only by accident. There could perhaps still be generation in the purely empirical sense of this term—assuming, of course, that such an empirical sense could be cleansed of all the metaphysical baggage just discussed, which is not at all obvious. Generation could perhaps still happen without this seed, but there could be no perception of this generation as generation without a prior relation to an indivisible immortality that births the desire for generation in the first place, and does so for plant, animal, and human alike. Plants, animals, and humans might still produce offspring. Life might still emerge. But according to Plato, by way of Diotima, without this prior relation to immortality, nothing guarantees that we or they could understand, perceive, or recognize anything like phenomena of production, offspring, or generation, let alone reproduction, life, parent, or child, even in the most basic, immediate, empirical, or medicoscientific senses of each of these terms. Even the most rudimentary, natural, or biological meaning of each of these givens is only provided through prior access to an immortality that instills them with their very sense. They contain, already, the kernel of immortality within.

Thus, biological reproduction, or what we call biological reproduction, although initially exemplary—for what more immediate, more natural, or more irrefutable example of the aim and essence of life could there be?—we learn in the passage above is itself only a figure for an immortality, with respect to which it is necessarily derivative, from which it draws its sense, and for whose sake it arises in the first place:

> Every mortal thing is preserved in this way; not by keeping it exactly the same for ever, like the divine, but by replacing what goes off or is antiquated with something fresh, in the semblance of the original. Through this device, Socrates, a mortal thing partakes of immortality, both in its body and in all other respects; by no other means can it be done. So do not wonder if everything naturally values its own offshoot. (208a–b)[28]

Pregnancy and reproduction are, in essence, secondary and supplemental. They are the devices, artifacts, contrivances, even ruses (μηχανῇ) of a mortal being. They are the artifactual prostheses of life's true form, mere semblances of originary, immortal being. They are therefore already symbolic, mere copies of an original Ur-life, whose essence we've simply forgotten and whose sense they allow us to recall.

Unlike for Aristophanes, for whom a literal, historical schism preoriented all future pathways to corporeal recovery, for Diotima, it is only because what we today call reproduction is already secondary, only because it is already symbolic and supplemental in its very essence, that it could in turn be replaced by other, apparently superior modes of participation in the divine, such as the construction of institutions and the composition of beautiful speeches (208c–209e). Thus, it is only because reproduction is already a ruse, fitted to extend a life that is by its nature finite but that wishes it were infinite, that this self is itself susceptible to unlimited and indefinite modes of autoextension, beyond any preconceivable limits, and be it by blood or fame. For Diotima, life is always already marked by forms of survival and undergirded by an originary immortality that sets it on its path. Human life, although definite with respect to its end (immortality) like all other life-forms, is nevertheless exceptional insofar as it remains indefinite with respect to its modes of participation in this end.[29]

To love, it should by now be clear, insofar as it is for Diotima always an expression of the desire for immortality, is also, and always, a mode of self-generation. Insofar as modes of generation are now understood to be categorically motivated by a desire for the immortalization of the life that is generating, they are by necessity always already folded back on that life. This would remain true no matter how disparate and disseminated its products may initially appear to be. To love, to compose, to produce, to procreate, to engender, to beget, are always already—if one adopts this view—modes of reproduction, self-reproduction, autoextension, and self-inscription in the other, but an other that is always already determined to be the same.[30] To love, which is to say *to be,* is to produce oneself in the other; it is to produce oneself as oneself, which is to participate in and simulate immortality, be it through the extension of one's stock or one's name, and regardless of one's actual awareness of this situation.[31] All divisions arise in the wake of and as a response to this fundamental structure of mortal nature. For Diotima, the self, insofar as it is, and be it animal, plant, or human, is just this (unconscious) drive to extend itself. "Generation" thus here names this movement of autoextension before it names anything like fecundity or birth.[32]

Whatever else we may think of it, the explanatory power of Diotima's discourse is remarkable. She is simultaneously able to account not only for the bond between the sexes (as well as the break) but also for the relation between the human and the animal, and the link, however obscure, between the everyday experience of love and the rarefied passion for philosophy. If Aristophanes's earlier account of human nature as traumatic loss initially prevailed on Agathon and the others with its own tragic beauty, then Diotima's thesis, proffered by Socrates, takes that which is most forceful in the former and advances it a step beyond. Channeling the pathos of Aristophanes's unrecoverable eros into a mobile—and mobilizable—energy, it turns the tears of romantic melancholy into the wide-eyed zealotry of philosophical militancy. Werther becomes Jesus, Alcibiades Socrates.

There is, of course, a catch, a price that one must pay for all this clarity, for the self-assurance of knowing that *knowing* is the path to future happiness, because one is confident that all forms of love—of giving birth in beauty—are ultimately not only continuous but also able to be hierarchized. Accordingly, the eventual disappointment must be equal to the original excitement. To the precise extent that Diotima's account offers an all-encompassing understanding of the multiform manifestations of life—both the flying and the grounded animals, both woman and man, both moneymaker and philosopher—it of course also risks diminishing just what it shows by unifying all of life's spectacular diversity through this single reductive principle.[33] Could that really be all life is? A merely stratified set of investments in the worst kind of narcissism? An only apparently diverse commitment to what is at bottom a mere reproductive futurism, which now encompasses both organic and inorganic productions, human and animal alike?[34]

If this is the threat of the Diotimean—or Platonic—vision of love; if in its rapacious appetite for understanding it risks the reduction of all living things to a single organizing principle; and if in doing so it appears too violent, because too efficient and too univocal, at least for anyone who has seen even a single episode of *Planet Earth,* then what at bottom authorizes the interpretation? What authorizes the interpretation of all of the variegated modes of γέννησις, which term encompasses not only the modern senses of "genesis," "engendering," and "procreation" but also "production" and "making"—everything, as I have already noted, from the wild animal's begetting of offspring, to the poet's composing of verse, to the lawmaker's engendering of "child" laws (209d)—as being essentially continuous, progressive activities, performed in the name of an immortality nowhere visible? What authorizes the interpretation, in short, of life, in all of its infinite variations, as not only productive but essentially always already reproductive—which is to say, as always already caught up in a process of reflexive autoextension, whether through the creation of a mimetic filiation by way of the living, breathing child, or a metonymic nominal association in an inanimate but no less jealous textual offspring? If all modes of production are ultimately performed in the name of an eternal life nowhere actualizable, then every mode of producing offspring would ultimately come down to the same principle of autoextension, and, if I may anticipate my argument a bit, life extension, now realized by other means. On what grounds could one dare to circumscribe the very possibility of production, birth, and creation, the δύναμις of the ζῷον, under the umbrella of a reproductivity that henceforth becomes destined to preorient it? Does this not amount to arguing that love, that what is called love, is in its essence appropriative, always part and parcel of a movement of gathering and self-identification?

Well, Diotima might respond, what authorizes such an interpretation, what compels one to determine the referent of mortal nature in immortality, and thus as an ongoing and never-ending process of self-reproduction (be it biological, onomastic, graphic, or other), is the danger that without it, precisely what one already thinks that one knows about the animal desire to bear offspring would itself become a total mystery, and the link between what we today would call production (as poiesis) and reproduction (as parturition) would be lost. Without the reference to immortality, in other words, without this implicit and essential structural exigency of mortal nature to participate in, share an element of, or possess by proxy what is called immortality, "both in its body and in all other respects" (208b), immortality, this ideal, atomic kernel of identity, selfsameness, and ideality in an ever-changing world, well, not only would we no longer be able to account for the human desire to make poems and found institutions, but we also would no longer be authorized even to read the most basic forms of animal engendering as reproductive or procreative.[35] Without immortality, both making and remaking, both production and reproduction, would suddenly become enigmatic because the common referent of each activity would be lost. Mortal nature, being as it is all change and mutability, would know nothing of self-presence or self-identity, and we would be forced to admit that the ends of life are as various as its activities in that they ultimately lack a final term able to consolidate them.[36] For Diotima, one could not think mortality without a prior reference to immortality, because it is this impossible-to-attain yet immemorially inscribed telos that preorients mortal nature in its very essence and renders it comprehensible. What is life, then, if not simply this impossible striving for an immortality that it cannot hope to attain, if not, precisely, an endless process of inventing new pathways to deathlessness? And what is philosophy if not simply the opening of life to knowledge about what makes life worth living, but also, the first (and final) step toward putting this knowledge to practice, toward not only knowing but also becoming immortal?

For Diotima, mortal life requires immortality in order to become thinkable, because at the limit, what the desire for immortality names is just that for immutable self-presence, for a mode of being that would be identical to itself, just as it is the privilege of the gods to possess. Without this common, shared trait, without the trace of immortality that is to be found in and as the structure of mortal nature, without this clear and simple opposition to death, there would cease to be any unifying link among living beings. Not only would be lost the end of life, as it is shared by all mortal natures, but also any understanding of the activity of the living, of the principle by which life orients itself in the here and now. To grasp mortal existence as a supplement of immortality is thus to render its multifarious productions understandable, from the simplest forms of animal reproduction to the highest forms of philosophic maieutics.

Coda

The above situation raises a number of questions regarding not only what might happen to life if one is forced to admit that this phantasm or lure of immortality, understood as immutable self-presence, is just such a phantasm, but also how the technology of the concept of immortality—how the concept of immortality understood as a technology and as an invention—is mobilized alongside so many others in order to shore up a conception of production (or generation) as reproduction. The concept of immortality is an invention. It is an invention of life. At the same time, the concept of immortality is not *one*. The very idea of the unicity of the concept of immortality is an effect of that conception, which invents itself—in Plato, for instance—as the very possibility of its unity. Immortality, we have just seen, grounds itself. To read immortality in the wake of the failure of its autopositing ground is to begin to witness its indefinite processes of virtualization. In Plato, it is mortal life's failure to attain immortal life that opens the indefinite process of prosthetic immortalization via unlimited, variegated means. These means cannot be anticipated. One cannot predict the directions they may take in advance. Yet the ground for them remains wholly circumscribed. Once the performative character of the concept of immortality (as one) is revealed, however, it becomes possible to bear witness to how its performative dimension does not simply reproduce the same life on an endless loop but participates in reinventing the very life that ought to preorient it. While we may never be able to wholly eliminate the effects of this performance, Plato's text allows us to observe and even to participate in its ongoing reinvention.

Immortality does not simply follow life. Immortality invents it, transforming that which will have been valued of life, interpreted as life, delimited as life, translated from life, in advance. Immortality is an originary translation of life that always precedes and follows it.

Diotima, in affirming the unicity of immortality, reveals the rhetorical nature of this performance, and in so doing, she opens the path to reading its signatures otherwise. Moving forward, we will see both how life produces itself as a process of prosthetic immortalization and how these prostheses always potentially risk transforming that life into something else.

2

DERRIDA AND THE MORTAL IMMORTALITY OF SOVEREIGNTY

> Sovereignty—the soul, and therefore the life of the state, the artificial respiration of the state—is posited, instituted, promised, contracted, *artificially, as immortal* only because it is *naturally mortal.* It is prosthetic and artificial technique that immortalizes it or in any case guarantees it an indefinite survival.
>
> —Jacques Derrida, *The Beast and the Sovereign*

In *The Beast and the Sovereign,* while examining the structure of sovereignty as it is set forth in the *Leviathan,* Derrida identifies a certain immortalizing tendency at work within the logic of Hobbes's political monster. According to the argument of the *Leviathan,* Derrida argues, immortality is both a requirement for the sovereign and an impossibility for it. The sovereign is, by necessity, posited as immortal. Insofar as its immortality must be posited, however, it is also invalidated in advance. To posit, effect, or perform immortality is to undermine that very performance, for the immortal, like the sovereign, must be, not become. Immortality, Derrida thus concludes, necessarily emerges as the effect of prosthesis, but to that extent, it is also undermined by its own artifactual, contractual structure. It is this situation, then, that ultimately leads him, in subsequent efforts to interpret Hobbes's example and to assess its generality, to refer to what he calls the "mortal immortality of sovereignty."[1] Sovereignty's immortality is, in the final analysis, a mortal one. Nevertheless, something within its fundamental finitude leads it to perform itself as what it is not.

What Derrida here identifies, at the end of his life, and by way of

Hobbes's text, is only one of many such examples of immortalizing logics that he articulates in his work. As we shall see, these examples return with some regularity, in diverse and proliferating fashions, and in several contexts. In each case, however, they subscribe to what I am calling, in the wake of Plato's example, the general logic of prosthetic immortalization. Within *The Beast and the Sovereign,* this logic consists in a tendency—evidently inherent to sovereignty, but not only to sovereignty—of the mortal to immortalize itself by way of artifactual or external supplements. I say inherent not to sovereignty alone because it exceeds this field in at least two fashions, which prove decisive. First, this tendency may be found outside what one traditionally recognizes as sovereignty's specifically political sphere. It is at work, for example, within the constitution of the subject, being nothing less than a general structural feature of ipseity. Second, far from an attribute of the sovereign, or even of the self, it is just this tendency that constitutes the essence of sovereignty itself—or, more precisely, that accounts for the tendency to sovereignty, which is constitutive of sovereignty's very essence.[2] Being that there is no sovereignty, the logic of prosthetic immortalization names the ineluctable impulsion to perform sovereignty as the effacement of mortality in immortality.

What Derrida provides through his analysis of the mortal immortality of sovereignty is thus an account of the phenomenon of immortality—the story of its emergence, let us say by way of anticipation, as a phantasm—which, he shows, is an immediate, incontrovertible consequence of the (finite) logic of the trace. That is, the finitude, finiteness, or inveterate divisibility of the sovereign self implies, requires, or even grounds, as a necessary possibility, the positing movement whereby unicity, indivisibility, and immortality are set forth. It is on the basis of this account of the phantasmatic appearance of sovereignty that a new interpretation both of the meaning of immortality and of the irreducibility of its lure will be made possible.

Thus, where Diotima, discussed in chapter 1, grounded mortal life in a more originary, immortal presence, Derrida locates a pulsion to immortality from the more originary, finite structure of the trace. In this respect, everything that follows can be read as Derrida's implicit response to and critique of the Diotimean vision. At the same time, if I have dared to speak of logics of prosthetic immortalization at work in both contexts, it is because, although Derrida will take his distance from Diotima's Platonism in specific and crucial ways, what each example nevertheless ultimately forces us to confront is the irreducibility of immortalization for mortal life.

In this chapter, I will thus be concerned with showing how Derrida does not simply depart from the Platonic paradigm but instead complicates it while demonstrating why it remains, to a certain extent, irrevocable and even irresistible. Though he displaces the basic premise of Diotima's vision by grounding the unicity and indivisibility of an alleged infinite, immortal presence, in the trace and in survival, nevertheless, the immortalizing prosthesis remains. To read Derrida after, but also with, Plato is thus to explore what of Diotima's vision may in fact be general and irreducible as a haunting, structural feature, now of *différance.*

Kantorowicz and the Double Body of Sovereignty

In session 2 of the first volume of *The Beast and the Sovereign,* Derrida addresses the structure of Hobbes's famed *Leviathan* twice in the span of three pages. As a prime example of the phantasm of state sovereignty, at once mortal and immortal, the *Leviathan* bears all the traces of the logic of the king's two bodies, as laid out by Kantorowicz. At the same time, as eventually becomes clear in the gradual working out of the structure of the theologicopolitical, by way of Celan, Benveniste, and Machiavelli, far from a character proper only to medieval and early modern political theory, as Kantorowicz had shown, this double-body, mortal–immortal structure of the king turns out to be the very meaning of sovereignty as such for Derrida. Why is that? Why does this structure prove inexorable? How does the general character of sovereignty, which must now be thought by way of the binary of mortal versus immortal, in turn transform our understanding of this opposition? If everywhere there is sovereignty there is the problem of mortality, then is it also the case that everywhere there is mortality, there is also the problem of sovereignty? If so, then neither figure would ever be wholly severable from or comprehensible without the other. These questions will guide us as we explore why the mortal–immortal structure of sovereignty should prove unavoidable for Derrida.

In his initial approach to the *Leviathan,* Derrida broaches the question of sovereignty's mortal immortality by way of the problem of positing. Positing is what the sovereign must do in order to assert its infallible, immutable, immortal being. At the same time, positing is what betrays the artifactual, contractual nature of sovereign being, thereby undermining its very performance. Far from an exception to sovereign power, this contradiction marks the situation of sovereignty as such. Indeed, it lies at the very heart of sovereignty, whose constitutional mortality births these simulacrum effects even as it undermines them. It is this play that Hobbes's text brings to center stage:

> If there is a prosthetic structure to the Leviathan as political animal or monster, this is because of its conventional, thetic, contractual structure. The opposition between *physis* and *nomos* (nature and law), as opposition between *physis* and *thesis* (nature and convention, or nature and positing), is here fully and decisively functional. It follows that law, sovereignty, the institution of the state are historical and always provisional, let's say deconstructible, essentially fragile or finite or mortal, even if sovereignty is *posited as immortal.* It is posited as immortal and indivisible precisely because it is mortal and divisible, contract or convention being destined to ensure for it what it does not have, or is not, naturally. So that if sovereignty is, as Hobbes says, the "Soule of the Common-wealth," this soul is an artificial, institutional, prosthetic and mortal soul.[3]

According to the logic of sovereignty, as it is articulated by Hobbes and then interpreted by Derrida, the immortality of the sovereign is posited, but as posited, it is already given up, and hence the paradox of the prosthesis. The prosthesis—or, as he had earlier called it, "prosthstatics"—of the state makes possible the appearance of immortality—it is instituted, posited, contracted—but in doing so, it renders impossible the realization of the very thing at which it had aimed, and not only in fact but also in principle.[4] Immortality of sovereignty is impossible. Nevertheless, somehow, the word, trace, signifier, or surface appearance of this artifice of immortality remains.

What, then, to make of this remainder? What to make of the afterlife of this phantasmatic figure of the artificially instituted and therefore in-advance-invalided immortality, which Derrida here identifies? By what necessity is it posited, again and again, and in the very act that undermines it? It is worth noting that it is not only Derrida but also Hobbes himself who had recognized the artificiality of the immortality that is posited. The artificiality and therefore invalidity of the immortality of the sovereign is part of its structure. What becomes notable in these passages is thus the fact that, though its impossibility is acknowledged, it remains present in intention. In the words of Hobbes:

> The obligation of subjects to the sovereign is understood to last as long, and no longer, than the power lasteth by which he is able to protect them. . . . The sovereignty is the soul of the commonwealth, which, once departed from the body, the members do no more receive their motion from it. The end of obedience is protection, which, wheresoever a man seeth it, either in his own or in another's sword, nature applies his obedience to it, and his endeavour to maintain it. And though sovereignty, in the intention of them that make it, be immortal, yet is it in its own nature, not only subject to violent death by foreign war, but also through the ignorance and passions of men it hath in it, from the very institution, many seeds of a natural mortality by intestine discord.[5]

Hobbes recognizes that the internal structure of sovereignty renders it fundamentally mortal. It is not only "subject to violent death by foreign war" but also destined to mortality "by intestine discord." Yet despite this structural mortality—"from the very institution"—an intention remains in those who make it to render it immortal. For Derrida, this (nonsovereign) response of sovereignty, whereby it posits its artifactual immortality from a position of natural, inveterate, incontrovertible mortality, is, far from accident or error, rather understood to be calculated and even inevitable. It is *because* of the structural mortality of sovereignty that its immortality must be posited. But in what, then, actually lies the necessity of sovereignty itself, and as a result its entire theatrics of artifice and immortalization? Why, in other words, continues to arise the very "intention of them that make it" for an (immortal) sovereignty that never was nor shall be? Could the immortal sovereign do other than arise? Finally, in what, if anything, lies the significance of this, its residual appearance, even in the wake of its acknowledged impossibility?

On the following page of the session, Derrida addresses this paradoxical situation for a second time. The contradiction between the intention of immortality and the structure of mortality is not one that requires resolution—it is not one that can be solved—because the former is itself the effect of the latter; it is the latter that is the cause of the former. The structure of mortality, we could say, begets institutions of immortality even if it also inscribes fatality into them:

> In the paragraph of *Leviathan* that I am going to read, I emphasize the lexicon of *protection,* where this lexicon and its logic explain the paradox of the mortal immortality of sovereignty. Sovereignty—the soul, and therefore the life of the state, the artificial respiration of the state—is posited, instituted, promised, contracted, *artificially, as immortal* only because it is *naturally mortal.* It is prosthetic and artificial technique that immortalizes it or in any case guarantees it an indefinite survival.[6]

As in the previous passage, Derrida here describes the relation between mortality and immortality as necessary. It is because the state is mortal and divisible, historical and provisional, fragile and deconstructible, it is only because it is "naturally mortal," that sovereignty is posited, instituted, promised, and contracted "as immortal." That the one is a necessary prerequisite for the other, in this passage, is evident. What is less evident—at least in the immediate context of the second session of *The Beast and the Sovereign*—is the precise order of causality, the specific necessity at work, in and between these oppositional terms. It is clear that the one requires the other. Nevertheless, by saying that the positing of immortality is "only because of" mortality, the present passage leaves it ambiguous as to whether this necessity inheres in every such positing or in every such mortality. The ramifications of this ambiguity are far reaching and have everything to do with the prospects of a sovereignty-less future.

In the former case, while every positing of immortality would require a state of mortality to precede it—no artifice of immortality without natural mortality—nevertheless, this would not imply that every state of mortality in fact posits its own immortalization. It gives us to think not only that there are states of mortality that do not posit immortality, but also that these states are not even haunted by the possibility of this (apparent) deficiency; they remain, somehow, exempt from this threat. One could imagine a mortal (sovereign) absolved from the lure of its proper immortality, a selfless sovereign uninvested in the fantastic performance of its own immortalization, or even in the fantasy of its unending, indestructible, or indivisible existence. But wherever there is positing of immortality, know that it follows from an underlying structural mortality.

In the second case, by contrast, we would have to conclude that wherever there is mortality, there is also positing of immortality—be it explicit or implicit. The structure of mortality itself would be sufficient to know that the lure of immortality is not far off, that the mortal (sovereign) is, by rights, irreducibly inconsolable in this regard—in other words, inconsolable with respect to its proper sovereignty. No mortality, therefore, may exist without artifice of immortality. This is an evidently much bleaker interpretation because according to this logic, try as one might, one would be all the same helpless to secure even the most benign manifestation of sovereignty (be it in the king, the self, or the subject) from these delusions of grandeur. Or perhaps it would be better to say that one would be all the same helpless to secure even the most benign king, self, or subject from the delusion of *their* sovereignty—as indivisible or immortal—and thus from some artifactual positing tasked with effacing their constitutional mortality/divisibility.

In the case of this latter, evidently more pessimistic, interpretation, there would remain the possibility of critique, or deconstruction, of this systemic flaw. It would remain perfectible, as Derrida was fond of saying.[7] Nevertheless, something inherent to the structure of sovereignty, which is to say to the structure of that which is not sovereign but wishes it were, would preclude ever eliminating this fascination as such. Sovereignty, in all its multitudinous configurations, would remain somehow uneducable as to the fantasy, and therefore also even feasibility, either of its immortalization or—which is not exactly the same thing, but which Derrida here links with the former—its indefinite survival.

Much rides on the choice between these interpretations. The attraction of the first, which is particularly strong today, is perhaps nowhere more visible than in the influential interpretation of Derrida's work put forward by Martin Hägglund. For Hägglund, the problem with traditional atheism is that while it rejects the existence of God and immortality, it nevertheless retains them as goods, finding nothing inherently wrong with them as ideals. Traditional atheism thus acknowledges the impossibility of these lures, but it retains them as lures. A radical atheism, by contrast—and more particularly still the kind of radical atheism to be found at work within Derrida's thought, argues Hägglund—rejects these ends not only in fact but also in principle, insofar as the conceptuality of each betrays a mode of being that would lie outside temporality and therefore outside the very finitude that makes any (desiring) subject possible in the first place. Thus, if Derrida, especially toward the end of his life, not infrequently referred to the perfectibility of institutions such as democracy, then for Hägglund, by contrast, one would have to speak of their correctability. That is, one would have to speak not only of the possibility of deconstructing these lures but also of eliminating them as lures. If, however, one accepts the second interpretation of the necessity of immortality's positing as one that inheres within mortality, then this would mean that the perpetuation of the lure—however contradictory and however antithetical to the structure that underlies it—would follow from what renders it impossible in the first place.[8] It would remain inveterate in spite of every intervention, and Diotima's vision, though metaphysical through and through, would remain to a certain extent irreducible.

I will argue that the thinking of *différance* in fact entails the second, evidently bleaker, and certainly less "radical" interpretation of these lines. Not only does every sovereign artifice of immortality presuppose an underlying mortality (what no one—not Derrida, Hägglund, or Heidegger—would deny), but also every naturally mortal sovereign also remains haunted by the phantasm of its immortalization. What must be attended to, therefore, is the difference between the multifarious projections of this phantasm, for example as indefinite or infinite, and the meaning of their apparent opposition. I treat this difference in Part II.

Ultimately, at stake in this interpretation are both an understanding of sovereignty and an understanding of mortality: of "the sovereign"—of every so-called sovereign—as mortal, and of "the mortal"—of every mortal qua mortal—as at least a little bit sovereign. This is because, as the example of Hobbes's *Leviathan* already shows, with its grand analogy between the individual and the state, or between the mortal subject and the mortal–immortal sovereign, there can be no rigorous division between these figures. It is the fact of their inextricability, with one always presupposing a conception of the other within itself, that Derrida is on the way to demonstrating in the seminars comprising *The Beast and the Sovereign;* it is, thematically speaking, that which can be seen to determine their trajectory. It is such a connection, or at least the possibility of such a connection, that determines Derrida's movement from a discussion of divine right and the authority of the state in the first year of the seminar to, in the second, a focus on questions of death, burial, and finitude by way of Heidegger and Defoe.

The radicalization of sovereignty here underway thus entails a certain convertibility between the self (as mortal, finite, and fragile) and the sovereign (as immortal, indivisible, and infallible) because for Derrida, and as shall be confirmed in a moment, there is no self without sovereignty, nor any sovereignty without self. This convertibility, moreover, will mean that what goes for the (so-called mortal) individual also goes for the (so-called immortal) sovereign, and even what goes for the mortal–immortal sovereign also and already goes for the mere mortal. Two important passages from subsequent sessions of the seminar's first year bring this point home. They show, on the one hand, that there is sovereignty—or at least a problem of sovereignty—everywhere that there is self, and, on the other hand, that the mortal immortality of Hobbes's *Leviathan* is endemic to all sovereignties in general, and therefore also even to the most finite finitude. Before returning to our guiding question—concerning the irreducibility of the positing of immortality and indivisibility for the naturally mortal and divisible subject or sovereign, as opposed to the mere irreducibility of structural mortality within every (therefore contingent) immortalizing prosthesis—it is worth putting these passages together to establish the scope of Derrida's demonstration.

The first such passage, concerning the generalizability of what Derrida calls the concept of sovereignty, is found in the immediately following session, in a discussion of Benveniste's findings vis-à-vis the etymology of ipse. Derrida draws on Benveniste, as usual, because the latter's uncovering of suggestive lexical resources aids him in the formalization of principles already observed to be at work within specific philosophical contexts—here, for example, those of Hobbes and Rousseau. By helping break down the ossified borders now separating our determinate linguistic usages, Benveniste's philological work allows for a broader glimpse into underlying structures of linguistic and philosophical conceptuality. Benveniste, in other words, both confirms and further complicates what Derrida had already suspected of Hobbes, and it is on the basis of Benveniste's writings on the long history of "mastery" that Derrida will, in turn, venture this early but decisive formulation of the formal structure of sovereignty in the seminar:

> The sovereign appeared most often in the masculine figure of the king, the master, the chief, the *paterfamilias,* or the husband—of the *ipseity* of the *ipse,* concerning which a few years ago, reading Benveniste, we had emphasized that, in its very etymology, it implies the exercise of power by someone it suffices to designate as *himself, ipse.* The sovereign, in the broadest sense of the term, is he who has the right and the strength to be and be recognized as *himself, the same, properly the same as himself.* Benveniste . . . goes on to wonder about the filiation of *poti,* Sanskrit *patyate,* Latin *potior* (to have power over something, to have at one's disposal, to possess, *potsedere*). . . . Now, and I insist on this point, for we will not cease to measure its consequences, Benveniste had inscribed the value of ipseity, the *ipse,* the "oneself," the "him(one)self" in the same filiation, as though power were first of all recognized in the one who could be designated or who could be the first to designate himself as *the same, a himself, a oneself.*[9]

For Derrida, following Benveniste, before there is the sovereignty of "the sovereign," there is the sovereignty of the self, he who (or that which) designates himself as *"the same, a himself, a oneself."* The decisive link between sovereign and subject, or sovereign and ipse, is therefore a power or capacity to say "I"—that is, to posit—before any endowment by city, state, nation, or church can come along and institutionalize the former in the body of a king or pope. This is why, already at the top of the quotation, it is a matter "most often [of] masculine" figures of sovereignty, in "the king, the master, the chief, the *paterfamilias,* or the husband—of the *ipseity* of the *ipse.*" Before there is sovereign as such, there is the "power" recognized of the one capable of designating himself as himself. Derrida continues:

> We commented a few years back on these lines of Benveniste, who is surprised that a word meaning "master" (as he had shown a page earlier: *potis,* in Sanskrit *patih* [master and husband], in Greek *posis* [husband], in composition *despotēs*) should become so enfeebled (Benveniste's word and evaluation, but I do not believe one can speak here of enfeeblement, of loss of power—to the contrary) as to signify "himself." On the other hand, Benveniste understands better that a word signifying "himself" could have taken on the proper sense <of> "master." . . . Plautus, then, uses the word *ipsissimus* (the *absolutely himself*) to designate the master, the boss, the most important personage, in short the first, the *princeps* or the prince. In short, the sovereign . . . among the Pythagoreans, the master Pythagoras was designated as *autos, autos epha:* he said it himself, i.e. the master said. . . . The master (and what is said of the master is easily transferable to the first of all, the prince, the sovereign), the master is he who is said to be, and who can say "himself" to be, the (self-)same, "myself."[10]

Across different languages, cultures, and historical contexts, the etymology of "master" bears out this link to the evidently exceptional capacity to say "oneself." Derrida will in effect link this minimal exception of the master who designates himself as such to the sovereign exception, recognized in the power of one to decide on the exception, as later formalized by Schmitt. Before, however, one can speak of any "state of exception," sovereignty will have come back to this exceptional capacity to say "I," on the very grounds of which one evidently separates between husband and wife, master and servant, sovereign and subject. Why does sovereignty come back to this capacity?

> The concept of sovereignty will always imply the possibility of this positionally, this thesis, this self-thesis, this autoposition of him who posits or posits himself as *ipse, the (self-)same, oneself.* And that will be just as much the case for all the "firsts," for the sovereign as princely person, the monarch or the emperor or the dictator, as for the people in a democracy, or even for the citizen-subject in the exercise of his sovereign liberty (for example, when he votes or places his secret ballot in the box,

> sovereignly). In sum, wherever there is a decision worthy of the name, in the classical sense of the term. Dictatorship (and in a minimal and strict sense sovereignty is always a moment of dictatorship, even if one does not live in a so-called dictatorial regime) is always the essence of sovereignty, where it is linked to the power to say in the form of dictation, prescription, order or *diktat*. From the Roman *dictatura*, where the *dictator* is the supreme and extraordinary magistrate, sometimes the first magistrate, and so the master of certain cities, to modern forms of dictatorship such as *Führer* or *Duce* or the Little Father of the People or some other "papadoc," but also in the figure of the dictatorship of the proletariat, in dictatorship in general as power that exercises itself unconditionally in the form of the *Diktat*, of the ultimate saying or the performative verdict that gives orders and has no account to render other than to itself *(ipse)*, and not to any superior agency, especially not a parliament—well, that dictatorship, that dictatorial agency, is at work everywhere, wherever there is sovereignty.[11]

The logic of sovereignty not only can but must be generalized in this way—extended all the way down, as it were—from the dictator to the chief, to the husband, to anyone who simply says "self" and decides—that is, to the minimal positing power to say "I."[12] This is not, as Derrida is careful to clarify, simply to equate these sovereign appeals, but rather to recognize the minimal working of the power to posit as a common thread linking them all. It is also to anticipate and refute in advance any facile effort simply to reject sovereignty in the name of a nonsovereign power as though one could refuse sovereignty—as though one could decide to refuse sovereignty—without doing so in its very name.[13]

The point of the above passage is, then, to show how participation in sovereignty adheres in originary, irreducible acts of positing before it can be a matter of specific, derivative, named institutions. As such, sovereignty cannot be isolated; it cannot be cordoned off or be made simply to oppose some nonsovereign alternative. One is always caught within certain relations of sovereignty. At the same time, this always already autoimmune sovereign expression is not uniformly manifested. There are degrees of sovereignty—degrees of autoimmunitary violence, indivisibility, self-presence, and self-mastery. Mortality and immortality—or fragility and immunity, temporality and atemporality, divisibility and unicity—here articulate the situation of sovereignty with respect to its own autopositionality, and they do so both historically speaking, as we shall see in a moment, and conceptually speaking, insofar as they repeat, for the subject, the dynamics at play in the theologicopolitical. Mortality (the sense of which, in the first volume of *The Beast and the Sovereign,* is not to be confused with finitude or being-toward-death, as Derrida will diagnose it in the second year of the seminar) here names the structural condition of divisibility that renders necessary, in the first place, the immunitary response that is (sovereign) positing (of sovereignty). "Positing" names at once the act of "the sovereign" and the act by which this sovereign may initially come to be as a sovereign self—hence the problematic status of the "auto" of autopositionality. But insofar as the sovereign's immunitary response must betray mortality, it is also already autoimmunitary. It is the firmest evidence of the failure of the act to affirm what it performatively enacts, and what it, in so doing, undermines. "Immortality" as the name for, if not the most extreme historical expression of, what in sovereignty ought to be absolute, indivisible, infinite, and infallible, is for this reason both always invalidated in advance and always aimed at, even when it is not aimed at by name. For this reason, it is no more subjective than it is constitutive of the subject. Like sovereignty, it adheres to the positionality of positing; it adheres to the very condition (mortality) whence positing arises in the first place.

From the third session, in which Derrida argues that there is sovereignty and thus some "dictatorial agency," some rejection of divisibility, fragility, or mortality, everywhere that there is positional logic of positing ("the master is he who is said to be, and who can say 'himself' to be, the (self-)same, 'myself' "), I now turn to session 11, in which Derrida stresses the irreducibility of the structure of mortal immortality for sovereignty as such.[14] Far from a structure exclusive to the *Leviathan*—or even to those medieval and early modern theories of political theology to which the *Leviathan* belongs—the logic of mortal immortality at work in Hobbes's text would be locatable everywhere that it is a matter of sovereignty, and thus to a certain extent also everywhere that there is subject, autopositioning of the I, or minimal articulation of the ipse. This is because what we call sovereignty and what we call mortality (or, now, mortal immortality) are in truth inextricable. The one always already implies the other.

Thus, it is in Derrida's eleventh session, nearly at the end of the first year of the seminar, that the question of mortal immortality returns, now through a staged confrontation between the historical figure of Louis le Grand on the one hand and the double-body structure of medieval political theology, as analyzed by Kantorowicz, on the other. It should be recalled that, according to Kantorowicz, the dual body structure of sovereignty was an integral element of medieval thought and political theory. Even as it created any number of juridical difficulties, it provided "an important heuristic fiction" for lawyers of the time, allowing them to reconcile modern with ancient law.[15] Although this was not acknowledged by the period's jurists, the notion supplies a prime example of a theological doctrine being translated into a juridical one. In the words of Kantorowicz:

> With the King's Two Bodies the [theological] doctrine of the *persona mixta* seems to have no direct relation. The duplication expressed by the concept of the *persona mixta* refers to temporal and spiritual capacities, but does not refer to bodies natural and politic. Or could it be that the king's impersonal and immortal super-body appeared, during the earlier Middle Ages, in some way or other, embedded in that very idea of his spiritual character resulting from the clericalization of the royal office? In fact, it is in that direction that Norman Anonymous, one of the staunchest defenders of the spiritual essence of a Christ-like kingship, sends us.[16]

In the *persona mixta* of Christian theology, one thus finds the early antecedent of the two-bodied king of English jurists. According to that earlier doctrine, both king and Christ are dual, or *geminatae,* insofar as they are both human as well as divine. The crucial difference between them, then, lies in the fashion by which they arrive at their mixed states: whereas Christ "was King and *Christus* by his very nature," "the king is a twinned being, human and divine . . . two-natured and geminate by grace only and within Time, and not by nature and (after the Ascension) within Eternity: the terrestrial king *is* not, he *becomes* a twin personality through his anointment and consecration."[17] Although such a distinction proves vital, allowing for an articulation of the difference between juridical and theological spheres, it is in fact the assurance of the continuity between them that guarantees the passage of power from God to the terrestrial world. As the eleventh-century Norman Anonymous writes, "The power of the king is the power of God. This power, namely, is God's by nature, and the king's by grace. Hence, the king, too, is God and Christ, but by grace; and whatsoever he does, he does not simply as a man, but as one who has become God and Christ by grace."[18] For the Anonymous, God, as prototype, model, and source of all earthly power, guarantees the power of the king. God endows king with his power through the privilege of his grace. Yet insofar as God is by nature immortal, indivisible, and all powerful, the problem for all future jurists will be how they reconcile sovereignty's inheritance of this divinity with the mortality, divisibility, and fragility endemic to every earthly frame. The double body of the king, as we shall see, responds to what for Derrida will become a structural problem for all sovereignty, insofar as all decision (in the classical sense of this term) involves some reference to this prototype—or rather, insofar as all decision always already performs the duality that the history of medieval political theology institutionalizes.

It is thus in light of this history that Derrida, arriving at the end of the first year of the seminar, again attempts to formalize in what consists the sovereignty of the sovereign. Turning to the famed scene of the Sun King's observation of an elephant necropsy, the question emerges for Derrida of the body of the king. This body, face to face with that of the dead elephant, sets into play all the operative oppositions of the year's discussions: dead and living, beast and sovereign, phenomenon and phenomenality. All stand at stake, side by side, in this duel. This situation leads Derrida to venture the following synthetic proposition concerning the divine monarch:

> In this dissection of the great elephant under the luminous gaze of Louis le Grand, the point was also no doubt to recognize in this great cadaver, right in the open body of a giant animal whose consent no one ever asked and who is at the mercy of the surgeons, of their scalpels, their gazes—[the point was also no doubt to recognize] in this great animal (great but not

> the king of the animals, the lion)—to recognize anatomo-physiological analogies: the brain, the heart, the kidneys, the circulation of the blood, etc.
>
> To reconstitute properly all these historical possibilities, we should have to go in many directions, well beyond what we would have time to do here. First, and minimally, into the analysis of what in this way makes visible an animal being, a *zoon* henceforth lifeless but previously alive, the submissive cadaver laid out, exposed, carved up, explored, the cadaver of one of the greatest, if not the greatest, animal in the world and, opposite, a little above it no doubt, the living body of a king who holds himself to be the greatest in the world; the living body of a king, then, but also the mortal body of a king who has at his disposal another body, that double body (earthly and mortal, and celestial, sublime, eternal, that of the function or the majesty, of the royal sovereignty supposed to survive eternally the mortal body—and sovereignty always presupposes this, in some way or another and in general, a priori, I would say, because it is in the structure of the concept of sovereignty, [sovereignty always presupposes] that double body the structure of which Kantorowicz analyzed in *The King's Two Bodies,* the double body that the Revolution naïvely thought it could end by decapitating the terrestrial body of Louis Capet.[19]

The duality of the king—of the mortal, disposable body of the king and the immortal, sublime body embodied in that body—Derrida "would say" is "presuppose[d]," "in some way or another and in general, a priori," by sovereignty. No simple matter of individual, historical, geographical, or cultural traditions or institutions of sovereignty, the double body, such as was analyzed by Kantorowicz, may be found at work everywhere—in general and a priori—that one finds sovereignty: "sovereignty always presupposes this," "that double body, the structure of which Kantorowicz analyzed," "it is in the structure of the concept of sovereignty."

How, then, are we to read this presupposition, or this necessity? Where to begin to address the ramifications of such a claim? What, in the final analysis, does it mean?

Something essential to the general structure of the concept of sovereignty lends itself to this double body, which we have just observed being played out in this scene, between the dead corpse of the elephant-animal and the living body of the king, but also within the body of that king, which is unto itself evidently already something both earthly and celestial, mortal and immortal, corporeal and sublime. The staging of Louis XIV's observation of the elephant necropsy plays out a drama—between living and dead, beast and sovereign, mortal and immortal—that, truth be told, one can already find at work within the body of Louis XIV himself. But that is not all, for in truth, what one observes in observing this observation is only the phenomenalization of a process—or, better, a movement—that, in point of fact, one can find at work wherever the haloed crown of sovereignty rears its brilliant head. This is to say that it exists wherever there is master, magistrate, dictate, dictator, or even ipse, ipseity, positing, or positionality of some "I." The structure of the concept of sovereignty is haunted by (the fiction of) this double body, which gains its power precisely by being transferable, be it by delving deeper and deeper into the autos of the ipse or by ascending farther and farther away from it, into structures persisting between organisms, subjects, colonies, or states. Sovereignty "is" as this fictive performance of difference that always comes down to the division between life and death.

If we now compare the a priori mortal–immortal double body of sovereignty to the mortal immortality of Hobbes's *Leviathan,* which posits its immortality only because it is mortal, then we may see that at stake in Louis XIV's observation of the elephant's corpse is just such a positing and just such a performance. Louis's observation (but also Derrida's observation of Louis's observation, and our own observation of Derrida's) phenomenalizes the opposition between sovereign and beast, living and dead, but also, therefore, immortal and mortal, through the "simulacrum-effect" that, Derrida points out on the following pages, is co-originary with sovereignty.[20] This simulacrum effect, or fiction, requires at bottom nothing more than the putting forth, or positional positing, of an "I" and its ob-ject.[21]

The Sun King becomes one through the presentation of this opposition, but now, instead of a performance local to the monarch, or even specific to the sovereign state as "political animal or monster," now, if one removes all the facts, so to speak, one may see that what is being played out is the process of prosthetic immortalization endemic to the minimal positing power of the "I," and one that may be said to happen in every such positing, insofar as positing is itself always already, and by necessity, the response to a condition for which the best name seems to be death. This double body, or mortal immortality, which, though naturally mortal, is not simply mortal—not simply content with its inveterate mortality—thus names a sort of obverse to "life death."[22] In other words, it announces the side of life as survival that cannot but aim at that second death that we call immortality, and that cannot but do so even if, on the one hand, it is destined to fail in this endeavor and, on the other hand, that at which it aims is in fact something that goes against its best interests.

The mortal immortality that Derrida lays out in *The Beast and the Sovereign* is something structural. It is the articulation of the rapport between the naturally mortal structure of the sovereign subject and the prosthetic performance, begotten by that structure, which sovereignly asserts its sovereignty by saying "I," and in so doing projects itself as such an unconditional, even immortal, power. Even before it can say either "mortal" or "immortal," the "I" speaks as a mortal, but what it says, evidently, underneath every enunciation, is just, "I am immortal."

Husserl and the Truth of Truth

As readers of Derrida's earlier, highly influential study of Husserl in *Speech and Phenomena* will by now recognize, there is something quite odd in all this. The logic that I have laid out, by way of the first year of *The Beast and the Sovereign* seminar, is without question related to that put forward some thirty-four years before, in the analysis of the role of iterability within the structure of the living present. Nevertheless, while there are clear lines of articulation running between these texts, one would hardly have thought to derive the above proposition from the arguments of the former, which appears to say nothing short of the opposite. It will be worthwhile, then, to look at a few critical passages from *Speech and Phenomena* to see how those arguments concerning the minimal conditions of possibility of the trace now relate to the simulacrum effects of sovereign and subject.

One of the most important insights of *Speech and Phenomena* concerned the logical necessity of the disparition of the subject for the intelligibility of any utterance. Just as in his subsequent treatment of

Condillac, in "Signature, Event, Context," where absence goes from a latent possibility of exclusively graphic writing to an originary, constitutive element of signification, Derrida here argues that, apropos of Husserl, the coming of death cannot be a secondary development that affects an already present subject. Rather, it must lie at that subject's very foundation. In a well-known passage concerning the subject and its inherent, incontrovertible, originary mortality, Derrida explains that according to the logic of Husserl's own argument, the appearance of the "I" requires the possibility of this subject's disappearance, and not as a mere contingency, but as an implicit, structuring condition of the emergence of the "I":

> If the possibility of my disappearance in general must somehow be experienced in order for a relationship with presence in general to be instituted, we can no longer say that the experience of the possibility of my absolute disappearance (my death) affects me, occurs to an *I am,* and modifies a subject. The *I am,* being experienced only as an *I am present,* itself presupposes the relationship with presence in general, with being as presence. The appearing of the *I* to itself in the *I am* is thus originally a relation with its own possible disappearance. Therefore, *I am* originally means *I am mortal. I am immortal* is an impossible proposition.[23]

The appearing of the "I," "of the *I* to itself in the *I am,*" we here learn, is "originally a relation with its own possible disappearance." This means, Derrida concludes, that "*I am* originally means *I am mortal. I am immortal* is an impossible proposition." "I am immortal" is, he explains, an impossible proposition. Yet the impossibility of this proposition appears to do nothing to counteract either the immortalizing prostheses of state institutions of sovereignty—as an intention, in the words of Hobbes—or the necessity, within the context of the philosophical demonstration of the principle of iterability, of reminding us of this proposition's very "impossibility"—which impossibility, we could then conclude, must not be entirely or immediately apparent. Impossible though it may be, it nevertheless remains, and insofar as it remains, it calls for a more nuanced account of its impossibility.

It is because things are more complicated, because the forces at work in the appearing of the "I" never simply come back to what is possible and what is impossible, that Derrida has affixed a footnote here in which he further clarifies the scope of this false, absurd, but nevertheless not meaningless proposition, which has had so much staying power. In what lies the necessity of this error, given its a priori absurdity, and be it at the level of the individual subject who says "I" or the state apparatus, which declares its indivisibility in perpetuity? How, in other words, are we to account for the appearing of the (impossible) double-body structure, within both political jurisprudence and the positional positing of the "I"?

> Employing distinctions from "pure logical grammar" and the *Formal and Transcendental Logic,* this impossibility must be expressed as follows: this proposition certainly makes sense, it constitutes intelligible speech, it is not *sinnlos;* but within this intelligibility and for the reason indicated, it is "absurd" (with the absurdity of contradiction—*Widersinnigkeit*) and *a fortiori* "false." But as the classical idea of truth, which guides these distinctions, has itself issued from such a concealment of the relationship with death, this "falsity" is the very truth of truth. Hence, it is in other completely different "categories" (if such thoughts can still be labeled thus) that these movements have to be interpreted.[24]

The "concealment of the relationship with death" is no less a part of the declaration of one's proper immortality than it is a portion of the classical idea of truth, Derrida here tells us. This is because both immortality and the idea of truth are based on the same presupposition: that of a pure, indivisible, grounding presence, and hence on the possibility of a transcendental subject that would be present to itself and not undermined in its very constitution through an originary relation to death. Once one recognizes the necessity of an originary relation to death, however, the proposition "I am immortal," though still intelligible, is revealed to be contradictory and necessarily false *(fausse)*. We could say that in its very appearing, it is revealed to be counterfeit. Nevertheless, because this "falsity" or even duplicity has its source in the evasion of death, it is also "the very truth of truth." It tells a truth higher than truth, a truth of truth no longer bounded by the constraints of the traditional concept of truth.

There are at least two ways we could interpret this truth of truth. On the one hand, the falsity of the proposition "I am immortal" is the truth of truth insofar as it reveals the lie at the heart of the traditional concept of truth. Every appearance of truth—in the classical sense of this word—appears only at the expense of a falsity that it cannot own up to. The truth of truth is, in this sense, false. On the other hand, the falsity of the proposition "I am immortal" is the "truth of truth" insofar as it tells the truth about truth. It tells a metatruth, one that requires recourse to a concept of truth no longer simply opposed to the false, and thus one that would appeal to "other categories," as Derrida here puts it, that would be up to the task of accounting for "such thoughts."

The proposition "I am immortal" tells the truth, in its very dishonesty, of the truth of truth. For this reason, Derrida calls for other categories, or at least for other terms, that would be capable of such thoughts. In concluding his note on this impossible proposition, he refers to the totality of what is at stake. This new language that he calls for would be capable not simply of revealing the falsity of propositions such as "I am immortal" but of interpreting what he now calls movements. That is, it would be capable of interpreting the process whereby the originary relation to death becomes effaced in the emergence of truth, of interpreting the movement whereby an impossible proposition such as "I am immortal"—which originally means "I am mortal"—can come about in the first place. How to account for such a movement? One that, like the counterfeit, which is neither simply true nor false, having to appear as what it is not in order to be what it is, would no longer make recourse to the simple opposition between truth and falsity? Derrida goes on:

> We can even go further: as a linguistic statement "I am he who am" is the admission of a mortal. The move which leads from the *I am* to the determination of my being as *res cogitans* (thus, as an immortality) is a move by which the origin of presence and ideality is concealed in the very presence and ideality it makes possible.[25]

What calls to be accounted for already in *Speech and Phenomena*—here where we learn that "I am immortal" is an impossible proposition—is nevertheless precisely the "movement" whereby "I am" determines my being as *res cogitans:* as immortality. This is, Derrida tells us, nothing other than "the movement" whereby "the origin of presence and ideality" *(différance)* is concealed in "the presence and ideality that it makes possible." "I am immortal" is an impossible proposition that is made possible by the very conditions that make it impossible: the structural mortality underlying it.

The mortality of the speaker, of the "I am," this "origin of presence and ideality," "makes possible" its own (self-)evasion. But here, just as in *The Beast and the Sovereign,* the reason for the necessity of this movement remains largely unaccounted for. In what lies the necessity of this slip, this movement, this shift, which leads to the "concealment of the relationship to death," be it for the classical idea of truth, the being of the sovereign, the mortal who says "I," or the *res cogitans?* Can one do anything but slip into the effacement of originary finitude? Can one do anything but strive for an immortality that is both impossible (as even Plato was ready to admit) and undesirable (as Hägglund has argued)? Can one do anything but this? It is no longer, on Derrida's account, an originary presence that gives mortality as its deficiency but rather the very structure of mortality—indeed, of survival—that would be productive of the teleological lures of origin and end as whole, selfsame, pure presences.

Unkillable Phantasms and the Kernel of Desire

From Hobbes to Husserl, from *The King's Two Bodies* to the mortal who says "I," and even from the sovereign to every self-pointing and thus self-positioning gesture that minimally indicates, in autodeictic fashion, a turn to me, one would thus have to locate the possibility of some structural sovereignty, which is to say the possibility for some phantasm of a contracted, counterfeit immortality within the mortal structure known as survival. At stake in our interpretation of the causality at play within the sovereign's mortal immortality is thus no less an understanding of the subject than it is one of life itself, to the extent that all life involves forms of autodeixis and autoaffection.[26]

Now, if sovereignty always presupposes this duality, then it can only be because no sovereign, or mortal subject, is ever wholly exempt from the threat of its own self-immortalization. This means that every subject, as mortal, in some sense presupposes the mortal immortality of Hobbes's *Leviathan*. It means that something internal to subjectivity, as it were, already implicates this structurally flawed and ill-fated process of prosthetic immortalization, even if only virtually. At least according to the progression of Derrida's thought as we have observed it in *The Beast and the Sovereign* and *Speech and Phenomena*, it appears that the mortal is *as* mortal immortal; the mortal subject is always already invested in its sovereign contractualization above and beyond the conditions of its finitude—not, of course, always in the same way or through the same media; not always explicitly and intentionally; and not necessarily in the mode of a dictator, a tech mogul, or a god, but rather in some way, be it materially or virtually, as an effect of the very trace-structure of life. Life contracts itself, it autoaffects itself, in and as the effacement of the trace that conditions it. Immortality, or what we call immortality, as well as what Derrida calls the sovereign's mortal immortality, I am arguing, is already contained in seed in this effacement.

If the generalizability of this movement of erasure and logic of prosthetic immortalization is easy enough to extrapolate from the passages examined above, it nevertheless remains unclear why it is that either should arise in the first place. Why, in other words, might something like sovereignty, or an effect of sovereignty, or a phantasm of sovereignty, as mortal immortality, be so ineluctably tied to positionality as to be generalizable in this way? Why is there something like sovereignty at all, in the first place, and thus effects of mortal immortality, or movements of effacement, for what is called the naturally mortal?

In order to understand why such a specter of sovereignty might come to be in the first place—which is to say why it is perpetually in the state of arising, again and again—and in order to understand why Derrida might be so adamant about Kantorowicz's two-body theory's being central to the general structure of sovereignty as such—and thus, by extension, to the (mortal) subject, which becomes subject only through recourse to a sovereign self-positing—it will be helpful to look beyond Derrida's late seminars, and even *Speech and Phenomena*, to other early interrogations of phonocentrism, the metaphysics of presence, and the phantasm. At least as early as *Of Grammatology* and *Writing and Difference*, the general structure of this problematic, within which the mortal sovereign's immortality would be but one example, comes into focus. There it is a question of the necessity of the historical denial of writing as trace in figures of speech and presence. How, Derrida asks, is one to account for this effacement or concealment? Ought it be considered accidental—and therefore correctible—or as ineluctable, because it is rooted in the very structural conditions that make it impossible in the first place? The interrogation of sovereignty, the question of the necessity of the sovereign's artificial, simulacral positings, and their relation to the structural (natural) mortality that underlies them, is just an extension of this earlier problematic, one in which the structural necessity of something like metaphysics is given account. This is the thread that will allow us to reconstitute the link between the mortal's sovereignty and the sovereign's mortality.

Perhaps nowhere in Derrida's work is there a more explicit account of the necessity of the movement detailed above, leading from the originary structure of the trace to the effacement of that origin as the very birth of truth and immortality, than in a little-known passage from *Ear of the Other.* The passage transcribes a response, given by Derrida, to a question posed by Rodolphe Gasché during a roundtable on translation. Spoken in Montreal after the presentation of "Otobiographies," Derrida's response is notable not only for its lucidity but also for the period in which it was given: in 1979, only a couple of years after the seminar on *Life Death,* and at the precise moment in his work that the motif of survival began to take hold.

Gasché asks Derrida three questions during the roundtable. They concern the relationship between Derrida's French and the French of the metropole, the problem of translation in Heidegger, and the significance of psychoanalysis for understanding interlinguistic translatability. In each of these questions, Derrida sees a common link, and that link is, precisely, the problem of commonality as he sees it at work within the figure of the "kernel": "These very difficult questions are very lucidly formulated, there where it is quite certain and clear that I cannot respond. Therefore, thank you. I am going to see if there is not among them some common kernel. And I believe there is one: it is, precisely, the 'kernel.' The question is whether there is a kernel intact somewhere or other."[27]

What follows is a lengthy commentary on the kernel and its relation to the structure of desire. After addressing the problem with Heidegger's understanding of Greek, Derrida turns his attention to the interplay between the "intact kernel" and "desire," or more precisely, to how the nonpresence of an intact kernel specifically produces the desire thereof. The argument goes as follows. The fact that there is no intact kernel is nothing less than *anankē,* or Necessity itself. There can be no intact kernel. Far from rendering moot the desire for such a kernel, however, it is its very nonexistence that births the desire for it. The desire for the intact kernel follows from that kernel's nonpresence. Moreover, it does so "irreducibly," "prehistorically," and "preoriginarily." It is the nature of this irreducible, prehistorical, and preoriginary relation to the intact kernel that interests us here. Although immortality is never named during the roundtable, what immortality names, for the Western tradition as well as for Derrida in *The Beast and the Sovereign,* is just such an indivisible, untouchable, virginal, intact kernel, the impossibility of which, we now learn, is precisely what spurs the desire for it.

Thus, although all desire is certainly not a desire for immortality (immortality is one name for the kernel at the heart of desire; there are others), what we must nevertheless come to terms with are the ramifications of an irreducible, prehistorical, and preoriginary relation to the impossible form of totality that immortality has (evidently) always named—or, anyway, has come to name in its Platonic–Cartesian–Hegelian lineage. If the relation to an intact kernel—an intact kernel that is not immortality but that carries the desire for immortality in seed—is irreducible, prehistorical, and preoriginary, as Derrida here argues, then desire for immortality is not simply one among others but also the secret desire of all desire, both bigger and smaller than desire itself. That which one desires in desiring (is) immortality (is) the secret desire of all desire: the intact kernel, just as Socrates once said. This does not, however, make Derrida a Platonist. Rather, it serves to reinscribe Platonism within the structure of the trace so as to preempt the priority of the kernel by a Necessity that makes its positing necessary in the first place. Although one is not fated to repeat Plato, one is nevertheless destined to maintain a relation to a certain lure, to the extent that this lure is itself structural, preoriginary, and prehistorical. Immortality has many names.

> The desire for the intact kernel is desire itself, which is to say that it is irreducible. There is a prehistoric, preoriginary relation to the intact kernel, and it is only beginning with this relation

> that any desire whatsoever can constitute itself. Thus, the desire or the *phantasm* of the intact kernel is irreducible—despite the fact that *there is no* intact kernel. I would oppose desire to necessity, to *anankē*. The *anankē* is that there is no intact kernel and there never has been one. That's what one wants to forget, and to forget that one has forgotten it. It's not that something has been forgotten; rather, one wants to forget that there is nothing to forget, that there has been nothing to forget. But one can only forget that there has never been an intact kernel. This phantasm, this desire for the intact kernel sets in motion every kind of desire, every kind of tongue, appeal, address. This is the necessity and it's a hard one, a terrible necessity. But just as without the desire for the intact kernel which doesn't exist, the desire for the untouchable, for virginity . . . just as without this desire for virginity no desire whatever would be set moving, likewise without Necessity and without what comes along to interrupt and thwart that desire, desire itself would not unfold. I don't know what else to call this but Necessity with a capital *N*, something that no one can do anything about and that is not a law instituted by any subject. Without this Necessity, it's law against law, desire against desire, proper name against proper name. But there is a law above these laws, which I am calling *anankē* and which controls it all (in this way, perhaps, I think in Greek, more in Greek than in Jewish). Thus, above the scene of the war between Yahweh and the Shemites, there is *anankē*, that is, a law which is not produced by any desire but which controls the struggle between these desires and these proper names. This *anankē*, no less than the desire for virginity, is what makes possible the kernel desire itself—the intact desire for intactness.[28]

For Derrida, the nonexistence of the intact kernel is incontrovertible. A law above the law, this Necessity sets into motion all subsequent desires, every kind of "tongue, appeal, address," prototypically among them "the intact desire for intactness." The lack of an intact kernel is the condition of possibility of desire as such. Nevertheless, what this lack conditions, in irreducible, preoriginary, and prehistorical fashion, is just a desire for its presence, which, were it realized, would of course undermine the very structure that makes it possible. The nonpresence of the kernel structurally grounds desire and destines it to desire its condition of impossibility in an untouched, virginal, intact kernel, whose absence we in turn wish to forget. Echoing this thought elsewhere, he asks:

> How could the desire for presence let itself be destroyed? It is desire itself. But what gives it, what gives it breath and necessity—what there is and what remains thus to be thought—is that which in the presence of the present does not present itself. Differance or the trace does not present itself, this almost nothing of the unpresentable is what philosophers always try to erase. It is this trace, however, that marks and relaunches all systems.[29]

Like the "forgetting" of *anankē,* what "philosophers always try to erase" is the "almost nothing of the unpresentable." In both cases, the unpresentable, differential condition of possibility of inscription begets forgetting and effacement. It begets a desire for the impossible itself, even though this goes against what "gives it breath." How could the desire for presence let itself be destroyed?[30]

With remarkable clarity, Derrida here lays out in the language of psychoanalysis the general structure that marks the metaphysics of presence—not only for the desiring subject but also for the sovereign and its immortality, as well as speech and its presence to itself. If we now compare the problem of sovereignty, as it is put forward in *The Beast and the Sovereign,* to the problem of the kernel, as it is articulated in this passage of *Ear of the Other,* we find a number of similarities. The same structure is at work in both scenes, if not the same desire. Common to both is the supposition of indivisibility: as virginal intactness and then as immortal imperviousness, the desire for which arises out of the originary impossibility of the indivisible. The fact that there is no indivisible, virginal, immortal origin sets loose the structure of desire as the birth of the phantasm.

In a moment, I will explore in what the particular importance of this phantasm consists. In other words, I will explore why the birth of the phantasm, as an expression of the desire for identity, indivisibility, and the intact kernel, cannot simply be dismissed as false or deluded, even if it names an impossibility. This phantasm, as Michael Naas has shown, always names for Derrida the privileged site of an irreducible appearance of identity, whose appearance, though simulacral, remains indestructible because it is based in the structure we are here outlining. Before turning my attention to this bizarre logic, however, let me first return to the initial passage quoted above from *The Beast and the Sovereign.*

Earlier, we saw that sovereignty "is posited as immortal and indivisible precisely because it is mortal and divisible, contract or convention being destined to ensure for it what it does not have, or is not, naturally."[31] Now, in *Ear of the Other,* Derrida identifies such a desire for the "intact kernel"—that is, the indivisible—to be "desire itself." The ground of such a desire for the intact kernel, for indivisibility—and immortality, I am arguing, is simply one more name for, one more trace of, or one more scene of the indivisible—well, this desire is now deemed to be irreducible because of a prehistoric, preoriginary relation to the intact kernel that is not.

In reading *The Beast and the Sovereign,* there was, at least initially, a question as to the precise order of causation between such an originary, "natural" mortality, and a posited, artificial immortality that sought to efface just that originary state. Here, the relation is evident: the nonexistence of the intact kernel not only makes possible the desire for this kernel (everywhere there is desire for intact kernel there is no intact kernel), but also, we now learn, this desire—which is "desire itself"—is itself irreducible, prehistorical, preoriginary (no absence of intact kernel without desire for it). The whole problem of positing, we could say, emerges out of this irreducible, preoriginary relation to a nonpresence that sets desire into play as a desire for presence, origin, and intactness. This is why, even were I to desire unintactness—or, analogously, mortality or finitude, the divisible structure of existence of the subject and sovereign—what Derrida calls at the end of the quotation the "intact desire for intactness" would in fact undermine this goal; it would betray that desire is always intact, even when it apparently aims at the unintact or at unintactness itself. To be completely, wholly unintact is also a version of intactness. Desire therefore always desires what is impossible for it. Desire is of the phantasm, in both senses of the genitive.

One can verify this in a number of fashions. It is implicit in Derrida's generalization of the logic of the king's two bodies to all forms of sovereignty, and in the subsequent radical extension of sovereignty to positing and positionality. But even without going to session 11, this is already intimated in session 3, when we learn that there is no simple contrary of sovereignty. Nonsovereignty, Derrida explains here, is also sovereignty. Every effort to oppose sovereignty, even in the name of the most democratic or anarchic state, participates in an element of sovereignty. This is not to reduce all governmental structures to a simple equivalence, or to underestimate the impact of political change, just as the acknowledgment that there is no pure and simple desire for mortality, without a trace of immortal lure, is not to level all theologies and atheologies. Rather, it is to acknowledge that intervention—and deconstructive intervention above all—does not destroy what it aims at but instead exposes the general laws within which such restricted economies operate. Deconstruction opens these discourses to what is exorbitant in them—to the incalculable—but only by maintaining a relation to calculation and by rendering the very difference between calculation and the incalculable undecidable:

> In a certain sense, there is no contrary of sovereignty, even if there are things other than sovereignty. Even in politics (and the question remains of knowing if the concept of sovereignty is political through and through)—even in politics, the choice is not between sovereignty and nonsovereignty, but among several forms of partings, partitions, divisions, conditions that come along to broach a sovereignty that is always supposed to be indivisible and unconditional. Whence the difficulty, awkwardness, aporia even, and the slowness, the always unequal development of such a deconstruction. This is less than ever the equivalent of a destruction. But recognizing that sovereignty is divisible, that it divides and partitions, even where there is any sovereignty left, is already to begin to deconstruct a pure concept of sovereignty that presupposes indivisibility.[32]

If the project of a deconstruction of sovereignty—or of immortality—cannot aim at "destroying" these lures or eliminating these phantasms, then it must aim instead at recognizing what has always been divisible within their concepts. The recognition of division, history, or partition within the phantasm is not sufficient to eliminate it. But it can transform it, and it can expose what desire cannot help but desire. The error would therefore be to take Derrida's call to dissociate "*unconditionality* and *sovereignty*" as proposing an alternative to this logic.[33] What he elsewhere calls the "revolution in the revolution," at work, for example, in Celan's *Meridian,* does not announce a break with sovereignty, mastery, or desire but puts us in contact with the more originary, presovereign conditions of possibility of sovereignty, mastery, and desire.[34] The affirmation of such unconditionality is called for by the logic of the event. Nevertheless, to take it as the memory of a premetaphysical, virginal origin without self-interest would be to risk repeating the metaphysical gesture by appealing to an outside of metaphysics.

Before turning, by way of conclusion, to places in Derrida's early writings where this preoriginary movement of desire and effacement is identified, I want to make a brief detour. I mentioned above that one key to understanding why immortality itself—the illusion of immortality as the phantasm of an itself—might be an irreducible, ineluctable movement rather than a merely contingent or preventable possibility, lies in Derrida's thinking of the phantasm. The phantasm stands out as a critical, though at times evasive, concept in his work. Although it appears to be proximate to the figures of ghost, phantom, specter, and spectrality, as Naas has shown, it in fact stands out from each of these terms in at least one important fashion that will prove critical for us. As Naas explains, while each of these other figures "suggest[s] a repeated or iterable appearance or apparition that calls into question or disrupts any notion of a stable, unchanging reality or identity behind or before those appearances," by contrast, "the notion of the phantasm in Derrida, while also always the *result* of a series of phenomenal iterations, nonetheless suggests some abiding reality or identity behind or before these phenomenal appearances."[35] In other words, whereas Derrida frequently calls on the ghost or specter to emphasize the bifid nature of an apparent presence, the special status of the phantasm is precisely that despite its in truth divided nature, it testifies to an abiding reality of this appearance (of indivisibility). Moreover, the phantasm forces one to confront the appearance of identity, of indivisibility, and of self-presence no longer as a simple fiction but instead as a fiction grounded in incontrovertible processes that must therefore be confronted on their own terms.

What is remarkable about the figure of the phantasm—but also, therefore, about the phantasm itself—is that it indicates a certain reality of appearance. It is not that that to which the phantasm refers—almost always identity, as Naas shows—is now somehow real, but rather that the appearance of identity that the phantasm performs has a certain reality, which calls for special account.[36] Though this reality calls for deconstruction, and exemplarily so, its specificity is to be undeconstructible, *increvable,* "unkillable," or insusceptible to dismantling.[37] One may try to kill the phantasm, but the phantasm's ultimate perversity lies in the fact that it is just this murderous tendency that keeps it going. As Derrida explains in "History of the Lie: Prolegomena,"

> As we know, *phantasma* also named for the Greeks the apparition of the specter, the vision of the phantom, or the phenomenon of the revenant. The fabulous and the phantasmatic have a feature in common: *stricto sensu,* in the classical and prevalent sense of these terms, they do not pertain to either the true or the false, the veracious or the mendacious. They are related, rather, to an irreducible species of the simulacrum or even of simulation, in the penumbral light of a virtuality that is neither being nor nothingness, nor even an order of the possible that an ontology or a mimetology could account for or subdue with reason. No more than myth, fable and phantasm are doubtless not truths or true statements as such, but neither are they errors or deceptions, false witnesses or perjuries.[38]

The phantasm is neither a truth nor simply a deception. It is related to "irreducible species of the simulacrum," and as such, it cannot be dismissed as error, perjury, or false witness. But then what does the phantasm serve to do? What is the phantasm of, in its not-simply-false simulation? As Pleshette DeArmitt writes:

> If Derrida's work, from the very beginning, put into question any putatively pure origin, pure autoaffection, pure presence, and any indivisible, inviolable center, the phantasm . . . is what leads us to believe in the coincidence of the self with itself, in immediate self-apprehension. The phantasm thus is "nothing

> other than our belief in a phenomenon that transcends itself, that spontaneously gives rise to itself." . . . The phantasm, it is important to underscore, would be "what our desire cannot fail to be tempted into believing."[39]

The phantasm is always the phantasm of identity. This quality of self-identity, or "self-coincidence," is prototypically identified by what we have already seen in Derrida's notion of ipseity. In the words of Naas, "The phantasm is always a phantasm of ipseity, a phantasm of self-identity and of the power generated from a being that is or is assumed to be identical to itself."[40] It is always "the phantasm of a self that can act, that has power, that is, in a word, *sovereign*." It is for this reason that

> to say that the self-coincidence or self-identity of a being is a phantasm is thus not at all to suggest that it is some powerless illusion. On the contrary—and this is no doubt one of the reasons why the phantasm so interested Derrida—the phantasm of ipseity is always a phantasm of power, the phantasm of a source or origin with the power to decide and to act. Even if such a source or origin is itself a phantasm, this phantasm is the source of all power and all sovereignty.[41]

Phantasm, sovereignty, ipseity—each of these figures would now be inseparable from the others. Each would condition the others, for in each case, the fictive performance of the phantasmatic presence would elide the conditions of that performance so that the phantasm may "present itself as self-generated, as naturally and purely given."[42] Insofar as it serves to cover up the conditions of its production, the phantasm is, Naas concludes, an essentially "reactive formation."[43]

The question that we have been addressing concerns, precisely, the necessity of this "reactive formation." It concerns the necessity of its staying power, be it as the "most unbreakable" of phantasms, as Derrida speaks of the "mother tongue," or as one of the "common phantasms," such as he considered narcissism.[44] Without doubt, the phantasm is the target of deconstruction: "the first impulse of what is called 'deconstruction' carries it toward this 'critique' of the phantasm or the axiom of purity . . . that would lead back to the indecomposable simplicity of the origin."[45] However, what the phenomenon of phantasms shows, time and again, from *Ear of the Other* to "The Night Watch," and like the sovereign's "mortal immortality," is a resilience that emerges from the conditions that render them impossible, as the temptation of the impossible.

The Irrepressible of *Différance*

I now turn, by way of conclusion, to a few places in Derrida's early writings where a similar account is given, though in slightly different terms. What, in *Ear of the Other,* is called a phantasm and is identified as being preoriginary, prehistorical, and irreducible is in *Of Grammatology* identified as a lure, and one that is already recognized to be indestructible. *Of Grammatology* is one of Derrida's earliest, most sustained interrogations of the metaphysics of presence and the systematic repression of writing. Thus, while one does not yet here find the phantasm in name, the dynamics governing the operation of *différance* nevertheless function according to the same logic: *différance,* this originary movement of differing and deferring, sets into play an indestructible lure of presence even as it contradicts it. *Différance* does not simply tolerate the effacement of its movement—the lure of or desire for its effacement—but gives rise to it, and it does so necessarily, indestructibly, systematically:

> Such would be the writing lesson in Jean-Jacques' existence. The act of writing would be essentially—and here in an exemplary fashion—the greatest sacrifice aiming at the greatest symbolic reappropriation of presence. From this point of view, Rousseau knew that death is not the simple outside of life. Death by writing also inaugurates life. . . . But the work of writing and the economy of differance will not be dominated by this classical conceptuality, this ontology, or this epistemology. On the contrary, these furnish its hidden premises. Differance does not *resist* appropriation, it does not impose an exterior limit upon it. Differance began by *broaching* alienation and it ends by leaving reappropriation *breached.* Until death. Death is the movement of differance to the extent that that movement is necessarily finite. This means that differance makes the opposition of presence and absence possible. Without the possibility of differance, the desire of presence as such would not find its

> breathing-space. That means by the same token that this desire carries in itself the destiny of its non-satisfaction. Differance produces what it forbids, makes possible the very thing that it makes impossible.[46]

If *différance* produces what it proscribes and if it makes possible just what it makes impossible, then it does so because it is finite. The finiteness of *différance,* the infinite finiteness of *différance,* destines it to just this unresolvable play.[47] It destines it to a certain, inveterate indefinition, one between the finite and infinite. One will never have closure concerning the finitude of *différance* because its finitude precludes even this totalization. What is indestructible is therefore not presence but that to which presence is the response, that which provokes and produces the lure of presence, over and over again, irrepressibly. The lure of presence, like the phantasm, is the (reactive) response to the originary nonpresence of *différance.* This response may be opened but never definitively closed.

We see this, once again, at the conclusion to Part I, in a discussion of the appearance of the proper name. This "appearance," Derrida clarifies, "is a necessary function."[48] Thus, like the phantasm, its lure, the lure of presence "in the absolute hearing-itself-speak," must "be *situated* as a function responding to an indestructible but relative necessity."[49] This amounts, according to Derrida, "to *situating* the metaphysics or the ontotheology of the logos."[50] While, then, "logocentrism" and "the metaphysics of presence" are historical and therefore deconstructible, the desire to which they respond is, precisely, irrepressible: "I have identified logocentrism and the metaphysics of presence as the exigent, powerful, systematic, and irrepressible desire for such a signified."[51]

Indivisible sovereign immortality, virginal intact kernel, absolute presence of the undifferentiated, literal, proper name—these now name many lures whose conditions of possibility and impossibility are to be found in *différance. Différance* gives birth to (the intact) desire (for intactness) while forever limiting its satisfactions. To deny the effectivity, the appearance, or the phantasmatic workings of any of these effects of *différance* would be to deny its very play because *différance* produces these lures of intactness that it also renders unrealizable. To call immortality undesirable, as Hägglund does, is for this reason to deny not only the most immediately evident experiential quality of desire but also to deny what, according to Derrida, would be the most incontrovertible, irrepressible, and preoriginary effect of the structure of *différance.* It is also therefore to risk falling into an "atheology" no less metaphysical than the theology against which it is directed. This risk is already identified, in the same year as *Of Grammatology,* at the conclusion to *Writing and Difference:*

> Just as there is a negative theology, there is a negative atheology. An accomplice of the former, it still pronounces the absence of a center, when it is play that should be affirmed. But is not the desire for a center, as a function of play itself, the indestructible itself? And in the repetition or return of play, how could the phantom of the center not call to us? It is here that the hesitation between writing as decentering and writing as an affirmation of play is infinite. This hesitation is part of play and links it to death. Hesitation occurs within a "who knows?" without subject or knowledge.[52]

Rather than an affirmation of life, to swear off the lure of immortality in the name of mortality would be akin to foreclosing the very play of *différance.* Instead of this, we must learn to analyze the structure of this lure as well as its relationship to what, today, is supposed to be simply opposed to it, in survival, finitude, amortality, and indefinite life.

PART II

FROM IMMORTAL TO INDEFINITE LIFE

Would it even be absurd to suppose this quality of melioration in the human species as susceptible of an indefinite advancement; to suppose that a period must one day arrive when death will be nothing more than the effect either of extraordinary accidents, or of the flow and gradual decay of the vital powers; and that the duration of the middle space, of the interval between the birth of man and this decay, will itself have no assignable limit? Certainly man will not become immortal; but may not the distance between the moment in which he draws his first breath, and the common term when, in the course of nature, without malady, without accident, he finds it impossible any longer to exist, be necessarily protracted? As we are now speaking of a progress that is capable of being represented with precision, by numerical quantities or by lines, we shall embrace the opportunity of explaining the two meanings that may be affixed to the word *indefinite.*

—Nicolas de Condorcet, *Outlines of an Historical View of the Progress of the Human Mind*

I cannot quit this subject without taking notice of these conjectures of Mr Godwin and Mr Condorcet, concerning the indefinite prolongation of human life, as a very curious instance of the longing of the soul after immortality.

—Thomas Robert Malthus, *An Essay on the Principle of Population*

These germ-cells, therefore, work against the death of the living substance and succeed in winning for it what we can only regard as potential immortality, though that may mean no more than a lengthening of the road to death.

It is only on this condition that the sexual function can prolong the cell's life and lend it the appearance of immortality.

—Sigmund Freud, *Beyond the Pleasure Principle*

It is one of the most remarkable things that in all of the biological sciences there is no clue as to the necessity of death. If you say we want to make perpetual motion, we have discovered enough laws as we studied physics to see that it is either absolutely impossible or else the laws are wrong. But there is nothing in biology yet found that indicates the inevitability of death. This suggests to me that it is not at all inevitable, and that it is only a matter of time before the biologists discover what it is that is causing us the trouble and that that terrible universal disease or temporariness of the human's body will be cured.

—Richard P. Feynman, "What Is and What Should Be the Role of Scientific Culture in Modern Society"

A key issue for cancer as well as for aging concerns telomere "beads," repeating sequences of DNA found at the end of chromosomes. Each time a cell reproduces, one bead drops off. Once a cell has reproduced to the point that all of its telomere beads have been expended, that cell is no longer able to divide and will die. If we could reverse this process, cells could survive indefinitely. Fortunately, recent research has found that only a single enzyme (telomerase) is needed to achieve this. The tricky part is to administer telomerase in such a way as not to cause cancer. Cancer cells possess a gene that produces telomerase, which effectively enables them to become immortal by reproducing indefinitely.

Molly 2004: *So I'll stay young indefinitely?*

Ray: *That's the idea.*

—Ray Kurzweil, *The Singularity Is Near*

> In one extreme scenario, the indefinite postponement of death will force societies to put severe constraints on the number of births allowed.
>
> —Francis Fukuyama, *Our Postmodern Future*

> If it is right and good to postpone death for a short while, it is difficult to see how it would not be better and more moral to postpone death for longer—even indefinitely.
>
> —John Harris, *Enhancing Evolution*

In opening our discussion of Derrida's diagnosis of the irrepressible, immortalizing tendency of sovereignty, we came across an equivocality in chapter 2 that we were not prepared to address. Reading Hobbes's *Leviathan* in *The Beast and the Sovereign,* Derrida had clarified that

> sovereignty—the soul, and therefore the life of the state, the artificial respiration of the state—is posited, instituted, promised, contracted, *artificially, as immortal* only because it is *naturally mortal.*[1]

Having thus established sovereignty's artifactual, prosthetic nature, and having linked this condition of inveterate mortality, divisibility, and historicity to the process of its positing *as* immortal, Derrida then adds, "It is prosthetic and artificial technique that immortalizes it or in any case guarantees it an indefinite survival."[2] The question was thereby raised as to how we are to understand the relationship between these two apparent alternatives: how to comprehend the as yet unresolved bond between immortalization and indefinite survival? Between the technic that immortalizes or, in any case, that guarantees *une survie indéfinie?*[3]

This is, as we shall see, an old question. It is one to which Diotima may already be said to respond—even if she does not do so explicitly—in her questioning of the relationship between generation and immortality. But it is also a question that, at least as concerns its unmistakably modern, biotechnological configuration, takes explicit form no later than the eighteenth century. It is a question, therefore, that forces one to ask it in multiple times, across multiple modalities. It appears no less pressing for the immortalist tradition that leads from Plato and Descartes to Hegel than it does for the mortalist tradition leading from Malthus up through Heidegger, Hägglund and Bostrom. Ultimately, however, and whichever of these traditions one pursues, one will be hard pressed to answer, in definitive fashion, in what this bond consists, and to establish, in clear and certain terms, the limit between the one and the other. Rather than a simple opposition, it is the link between each term, their irresolvable tension, that maintains and gives life to each. The decision between the technic that immortalizes and the technic that guarantees indefinite survival is undecidable.

Within the immediate context of *The Beast and the Sovereign,* Derrida's own apposition is less explanatory than it is suggestive. The polyvalence of his "or" does not let one decide but instead posits and even propagates a certain indecisiveness. It is not clear, for example, whether "indefinite survival" is here meant to serve as a modification of "immortality" or as a mere embellishment thereof, whether it is offered as a correction or simply as an appendix. Could it be that Derrida offers indefinite survival as an interpretation of immortality? But is it not just as likely that he proffers it as a hedge? At the same time, could we even say with certainty that Derrida here speaks in his own voice? Might he not be speaking for Hobbes when he speaks of immortalization, and only for himself when he speaks of indefinite survival? Is the takeaway that one must always speak of indefinite survival when one speaks of immortality? Or is it only that this is necessary within certain, specific contexts? But this raises yet another question. In what, exactly, lies the necessity of proffering both figures, which clearly bear some relation to one another, though not one that is fleshed out here (or anywhere else) in Derrida's work? Why offer both instead of one or the other? Are they simply interchangeable, and if not, what regulates the economy of their exchange? Is it even warranted to speak of them together in this way, in the same breath? Does one not already presuppose something about them when one associates them in this way? Would it not be better, therefore, to hold them apart, as Condorcet, but also Hägglund and Bostrom, would certainly advise?

To interpret the situation of this either/or is to confront difficulties both internal and external to the Derridean text. Although Derrida himself does little to clarify their relation, before and beyond the inscription of this equivocality within *The Beast and the Sovereign,* problems inherent to the history and conceptuality of each operative term force this pairing on his reading of Hobbes. The fact of this pairing's imposition—but also confusion—is a visible trait of their entire history, from Godwin up through Malthus, Freud, and Kurzweil, but also already in Plato and Aristotle, to say nothing of Shakespeare and Jacob. During this time, one has not ceased either opposing or comparing each figure to the other. If Derrida's text is for us exemplary, then it is only to the extent that it announces this problematic and exposes it, all while allowing it to remain suspended, unresolved, and in tension.

What is the relationship between immortality and indefinite survival, but also between indefinite survival and what we might be tempted to call definite survival, the so-called mortal state of finite finitude of which we, all of us, evidently already partake? From the perspective of Diotima, we would have to admit, along with Malthus, that each is simply an alibi for the other: that every movement of life, and above all those that work to extend it—but what movement is not also already in principle an extension?—do so in the name of an immortality nowhere present but everywhere felt. From Derrida's point of view, on the basis of what he has diagnosed as the quasitranscendentality of sovereign positing, it also appears that immortality is something endemic and irreducible, even if it nevertheless remains to be determined in what the precise relationship between each of these forms consists. It remains to be determined not only for what is called the human or even the posthuman but also for all of life, from the simplest, single-cell prokaryotic organism to viruses, colonial sponges, and silicon-based artificial intelligence. Where does sovereign positing begin? Where does it end?

In what follows, our question will initially be how to understand the relationship between the formal structures of mortality, immortality, and indefinite life. As we shall see, such a question also necessarily entails asking about the comportment of one who lives in the mode of each. Is there a difference between surviving immortally, as a mortal, and indefinitely? How to relate the one to the other? The purpose of Part II will be to give conceptual clarity to each of these terms as well as to the relations that hold between them. Such a project does not simply involve disentangling the one from the other, however. For essential reasons, when it comes to immortality and indefinite survival, even if one can establish relatively clear dividing lines conceptually, just as soon as it is a matter of positing one or the other—that is, of addressing the impulse to the latter or the former—something magnetizes each member of this pair to the other. If Plato and Derrida have given accounts of this complication, each in his own terms, then it remains to be seen whether and to what extent such accounts remain pertinent for our own epoch, notably for insights into the contemporary biological and biomedical sciences. It is this topic that I will address at length in Part III as we delve into the specific mechanisms by which biological life-forms can be said to live and even to be immortal. However, before turning to this investigation, or its implications for transhumanist projects of radical life extension, we must first gain conceptual clarity as to the meaning of each of these figures and the grounds on which they have been distinguished historically.

3

SUBSTANCE IMMORTALITY

Let us start with a preliminary distinction. It seems that immortality, in its classical, Christian conception, is understood to be both unconditional by nature and eternal by duration. This unconditionality and eternality may be interpreted either materially, as a certain indestructibility of substance, or ontologically, as a certain privation of access to time and being. If interpreted materially, then it can lead to certain forms of Cartesianism, and if interpreted ontologically then to certain forms of Heideggerianism. One commonly conflates these modes of interpretation today, and it would be fruitless to attempt to sever them completely, for reasons that I will discuss below. Nevertheless, each interpretation emphasizes a different valence of what is called mortality. On the one hand, mortality is understood as a susceptibility to and even as the inevitability of physical decay and destruction, which are inherent to all things material. On the other hand, mortality is taken as the experience of the impossible possibility of death, as given by originary temporality, or what is also known as phenomenological finitude or *Endlichkeit*.

The notion of immortality's unconditionality is particularly significant for its material interpretation, which views it as culminating in the being of an immutable substance. That immortality should be unconditional means that once someone or something becomes immortal, the process is irreversible. The necessity of its irreversibility is seen when the converse holds, for were the immortality of the immortal to be reversible—or revocable—even in principle, then this so-called immortal would turn out to be in fact mortal, and thus to have been mortal all along. True immortality, rather than its mere semblance, is not simply the impossibility of death at any given moment. Rather, it is the infinite extension of this impossibility, be it inside or outside time. The immortality of the most immortal immortal must therefore be both absolute and without condition, for only in this way can true indestructibility, or indestructibility *sans phrase,* be assured.

Nothing, then, including of an act of God—the immortal par excellence—ought to be capable of stripping the immortal of its immortality and subjecting it to change, finitude, temporality, annihilation, or death. But even if God could strip others of their immortality, nevertheless, nothing could strip him of his. The immortality of God is, in this sense, the most closed, infallible, and safe of all. Being the origin of all creation, God is immune to every externality and impervious to any and all outside force. His resistance to death is thus without conditions, be they physical or metaphysical, and is therefore both absolute and infinite. As Descartes writes in his *Meditations:*

> Sovereign God [is] eternal, infinite, immutable, all knowing [and] all powerful.[1]

> By the name God I understand a substance [that is] infinite, eternal, immutable, independent, all knowing [and] all powerful.[2]

> The unity, simplicity, or inseparability of everything that is in God is one of the principal perfections that I conceive to be in him.[3]

The God of Descartes is such that it is not even conceivable that he could become mortal, which is perhaps why, strictly speaking, he may not be immortal either. Descartes never actually refers to God in this way, at least not *à la lettre*—according, that is, to the letter of the *Discourse on Method* or *Meditations on First Philosophy.* Descartes never actually calls God immortal *(immortel, immortalem)* in either work, in either French or Latin. The very idea of his infinite being dictates his "unity" and "simplicity," and hence his a priori indivisibility and self-identity, from which perspective it follows that he must be eternal and unchanging, but perhaps not immortal, if only because this term still appears to imply that its subject holds a latent relation to mortality. The immortality of God, we might say, goes without saying. It is so absolute as to be ineffable. It is an immortality that effaces every relation to its possible negation.

That God must be eternal and immutable follows, for Descartes, as the immediate consequence of his infinite perfection, for such perfection, he goes on to explain, is without limit. God is not infinite merely in one way or another. He is not perfect simply in this way or that. His perfection is infinite and absolute, and this means that it is by necessity without limit: "There is nothing that I name properly infinite, except for that in which no limit can be found in any of the parts, in which sense God alone is infinite. . . . [The infinite possesses] no limitation."[4]

It is thus via analogy with the immutable, infinite perfection of God that one arrives at the inference of the soul's immortality as a pure—though not self-conditioned—substance. Man is of course most famously for Descartes a "composite" of soul and body. Man is an "I think" as well as an "I feel." Nevertheless, although one always finds oneself strictly "joined," "intermingled," and even "confounded" by these two dueling substances, the essentially perishable nature of the body does not ultimately detract from the essentially imperishable nature of the soul:[5]

> The human body, insofar as it differs from other bodies, is only formed and composed of a certain configuration of members and other similar accidents; and the human soul, to the contrary, is not at all composed of any accidents, but is a pure substance. Because even if all its accidents are changed, for example, if it conceives of certain things, if it wants other ones, if it senses others, etc., it is nevertheless always the same soul; whereas the human body is no longer the same as soon as the look of some of its parts find themselves changed. From which it follows that the human body can easily perish, but that the spirit, or the soul of man (between which I do not distinguish) is immortal by nature.[6]

The body's pathological divisibility—"we conceive any body only as divisible"—is thus directly contradicted by the soul's inveterate indivisibility—"the spirit or soul of man can only be conceived as indivisible."[7] On these grounds, the soul is determined to be immortal while the body mortal. Nevertheless, because the soul remains conditioned by God, Descartes cannot quite discount the possibility that God could one day rescind its privilege. The possibility of such a withdrawal—though remote—thus remains in his writings, thereby placing an asterisk over the only immortality that he properly identifies in the *Meditations:*

> Natural learning teaches us that the spirit is different from the body, and that it is a substance; and also that the human body, insofar as it differs from other bodies, is only composed of a certain configuration of members, and other similar accidents; and finally that the death of the body depends only on some division or change of figure. But we have neither any argument nor any example that persuades us that the death, or the annihilation of a substance such as the spirit is, must follow from a cause so slight as a change in figure, which is nothing but a mode, and even a mode not of the spirit, but of the body, which is genuinely distinct from the spirit. And, moreover, we have neither argument nor example, which can persuade us, that there are substances that are subject to being annihilated. Which suffices to conclude that the spirit, or the soul of man, to the extent that this may be known by the natural philosopher, is immortal.
>
> But if one asks whether God, by absolute power, hasn't perhaps determined that human souls cease to be at the same moment that the bodies to which they are united are destroyed, it is for God alone to respond.[8]

Could we not then say that immortality, for Descartes, remains prosthetic through and through? That it is, in other words, determined as a divine technology given—and therefore revocable—by God? That there is neither any natural nor pure immortality but only the contingent supplement, forever mediated through the infinite perfection of the all-powerful deity?

As Derrida points out, in speaking of the immortalizing sovereignty that is modeled after this Christian God, sovereignty "is always supposed to be indivisible and unconditional."[9] Being indivisible and unconditional, there is nothing to which it can be subjected, and nothing, then, capable of subjecting it to accident, event, or division. True substance immortality, then—or, in any event, the limit case of substance immortality, in the infinite perfection of (sovereign) God—is therefore not simply the possibility or the appearance of such indestructibility but the unconditional necessity thereof. This is why its most extreme manifestation is not to be found in the (perhaps revocable) immortality of souls, which are created by the sovereign and thus still subject to his will, but only in God himself (and this, even if it must therefore go without saying).

The realm in which this unconditional necessity becomes most fully expressed, for the Christian tradition that is Descartes's, is that of time: "First of all, because I cannot conceive of anything but God alone, to whose essence existence belongs with necessity. Then also, because it is not possible for me to conceive of two or more Gods in the same way. And, given that there is one now that exists, I see clearly that it is necessary that he had existed before all eternity, and that he will be eternally in the future."[10] For earlier, medieval theologians, such as Boethius and Augustine, although the thought of God's eternity was a clear conceptual necessity, it had nevertheless also been a puzzle. As the creator of all that is, including time, the eternity of God could not simply be conceived as a matter of duration, or everlasting and forever, but also before time, in the space that precedes time's creation, or as a form of timelessness. "Now that God is eternal," writes Boethius, "is the common judgement of all who live by reason."[11] He goes on:

> Therefore let us consider, what is eternity; for this makes plain to us both the divine nature and the divine knowledge. Eternity, then, is the whole, simultaneous and perfect possession of boundless life, which becomes clearer by comparison with temporal things. For whatever lives in time proceeds in the present from the past into the future, and there is nothing established in time which can embrace the whole space of its life equally, but tomorrow surely it does not yet grasp, while yesterday it has already lost. And in this day to day life you live no more than in that moving and transitory moment. . . . Whatever therefore comprehends and possesses at once the whole fullness of boundless life, and is such that neither is anything future lacking from it, nor has anything past flowed away, that is

> rightly held to be eternal, and that must necessarily both always be present to itself, possessing itself in the present, and hold as present the infinity of moving time.
>
> And therefore those are not right who, when they hear that Plato thought this world neither had a beginning in time nor would have an end, think that in this way the created world is made co-eternal with the Creator. For it is one thing to be drawn out through a life without bounds, which is what Plato attributes to the world, but it is a different thing to have embraced at once the whole presence of boundless life, which it is clear is the property of the divine mind.[12]

Where Plato conceives of "a life without bounds"—and in so doing thinks eternity still as essentially temporal—it is the distinction, argues Boethius, of God's eternity to "hold as present the infinity of moving time" and in this way to be distinctly atemporal. Similarly, writes Augustine in his *Confessions,* "it is not in time that you precede times."

> Otherwise you would not precede all times. In the sublimity of an eternity which is always in the present, you are before all things past and transcend all things future, because they are still to come, and when they have come they are past. . . . Your "years" neither go nor come. Ours come and go so that all may come in succession. All your "years" subsist in simultaneity, because they do not change; those going away are not thrust out by those coming in. . . . Your Today is eternity. . . . You created all times and you exist before all times. Nor was there any time when time did not exist.[13]

In the passage cited above from the *Meditations,* Descartes had appealed both to the figure of temporal eternity (or everlastingness) and to that of atemporal eternity (something like Boethius's whole presence, or Augustine's simultaneity). Nevertheless, for most, though not all, recent readers of Descartes, his God is understood to reside primarily in the latter, atemporal realm—a realm that one could not comprehend on the model of duration alone, even if God also has infinite extension within duration.[14] For our purposes, it is ultimately less important where one comes down within this debate—that is, how one understands God's or the soul's *aeternum,* whether as being inside or outside time, and the possible modifications within each of these positions are many—than the fact that in both cases one determines the immortal God as being infinite, which is to say without limit, beginning, or end. Both determinations of eternity, in other words, give expression to the degree of absoluteness of the indestructibility of the immortal substance—be it God or the soul—for Christianity. As we shall see, in doing so, they also allow for the transition from the first determination of immortality (as enduring substance) to the second (as mode of being, in the privation of temporality).

In *The Human Condition,* Arendt famously argues that the immortality of the Greeks was distinctly temporal in nature, while that of Christianity was distinctly atemporal.[15] Whether or not one follows her in this, in either case, to conceive of immortality as immutability (or imperviousness to death) throughout duration, or as immutability (or imperviousness to death) for all of timeless eternity, is likewise to remain invested in it as a mode of perdurance. Substance immortality is always immunitary in nature, invested in being as a mode of presence. It is a condition that one lives absolutely, unconditionally, and endlessly. And it is for this reason that, according to both the phenomenological tradition, from Heidegger on, and the biological sciences, as they have been constituted for at least the last hundred years, substance immortality implicitly contradicts the basic structure of life.

4

ONTOLOGICAL IMMORTALITY

> Imbedded in a cosmos where everything was immortal, mortality became the hallmark of human existence. Men are "the mortals," the only mortal thing in existence, because unlike animals they do not exist only as members of a species whose immortal life is guaranteed through procreation. The mortality of men lies in the fact that individual life, with a recognizable life-story from birth to death, rises out of biological life.
>
> —Hannah Arendt, *The Human Condition*

For Heidegger, in contrast to Descartes and Augustine, both of the above interpretations of eternity would remain bound to vulgar conceptions of time. To both must thus be contrasted an approach to immortality that interprets time neither as an infinite succession of present moments nor as the simple negation of this succession in an anterior, atemporal realm, but instead through Dasein's originary relation to temporality, which opens it to alterity and difference, as well as to the very possibility of negating this difference in experiences of eternity.[1] From the perspective of this latter thinking, then, the whole notion of indestructibility, so essential to immortality's material (or Cartesian) interpretation, would represent a mere effacement and forgetting—an effacement, that is, of this more originary relation, its closure, and even its foreclosure. For this latter thinking, to be immutable and selfsame in this way would mean rather to be without temporality—in the ontological sense of the word—and thus contrary to what a Descartes or a Plato might give us to think it would mean to be without life.[2] Indeed, according to the intellectual legacy of this tradition of phenomenological thought, if something or someone could in fact be impervious to death, could be eternal and immutable, it would, for this very reason, actually have to be dead, for only that which is dead cannot die.

As Hägglund has argued concerning such a Christian–Cartesian conception, if immortality and the desire for it involves an elimination of contingency, temporality, and conditionality, then it must in fact be equivalent to death—death being the true name for the state of absolute immunity.[3] If to be immortal is to be infallible and insusceptible to risk, then so too must it be without feeling or experience—which is to say without either affectivity or autoaffectivity, and thus without life. Derrida himself likens immortality to death in this way on several occasions. In *Given Time,* for example, he explains that "neither death nor immortal life can ever give anything, only a singular *surviving* can give," thereby implying a certain structural homogeneity between these states.[4] A similar sentiment is expressed years earlier, in *Of Grammatology,* when he diagnoses the common tendency of dualism and monism:

> All dualisms, all theories of the immortality of the soul or of the spirit, as well as all monisms, spiritualist or materialist, dialectical or vulgar, are the unique theme of a metaphysics whose entire history was compelled to strive toward the reduction of the trace. The subordination of the trace to the full presence summed up in the logos, the humbling of writing beneath a speech dreaming its plenitude, such are the gestures required by an onto-theology determining the archeological and eschatological meaning of being as presence, as Parousia, as life without difference: another name for death, historical metonymy where God's name holds death in check. That is why, if this movement begins its era in the form of Platonism, it ends in infinitist metaphysics. Only infinite being can reduce the difference in presence. In that sense, the name of God, at least as it is pronounced within classical rationalism, is the name of indifference itself.[5]

Life without difference: another name for death. Life without difference, absolute life, Derrida argues, is another name for death, and this is what the name of God metonymically captures, even as it evidently also serves to hold this death in check.

For Derrida, as for Hägglund, immortal life, in its Cartesian interpretation, must be equivalent to death, because said life involves an effacement of chance and therefore an absolute compression of life with death, rather than a sustaining of the differential rapport between them. Jean-Luc Nancy explores the implications of such a thought for human culture. For Nancy, the absolute presence at work in Descartes's vision of immortality corresponds to an "immanence" that also marks the metaphysical conception of "communion." By contrast, "community" names a bond that holds only between those who live by virtue of a relation to death:

> The genuine community of mortal beings, or death as community, establishes their impossible communion. Community therefore occupies a singular place: it assumes the impossibility of its own immanence, the impossibility of a communitarian being in the form of a subject. In a certain sense community acknowledges and inscribes—this is its peculiar gesture—the impossibility of community. . . . A community is the presentation to its members of their mortal truth (which amounts to saying that there is no community of immortal beings: one can imagine either a society or a communion of immortal beings, but not a community). Community is the presentation of the finitude and the irredeemable excess that make up finite being.[6]

"There is no community of immortal beings," explains Nancy, though there may be—or we can imagine there being—a "society or communion" of them.[7] In other words, the elimination of death as the finite being's proper, impossible possibility has as its consequence the elimination of what allows this individual to relate, in the first place, to the other in their specificity—a certain impossibility of communion. Without death (as presentation of finitude), there is therefore neither community nor a finite being as such. In the words of Christopher Fynsk, "The individual Dasein first knows community when it experiences the impossibility of communion or immanence (the self-presence of individuals to one another in and by their community) before the dead other."[8] Thus, while community requires the nonpresence of finitude, immortality, by contrast, entails self-presence, and it thus may allow for society or communion, but not for community in the full sense of this term.

Similar notions to Nancy's and Heidegger's can be found in Judith Butler's idea of precarity and Todd May's concept of fragility.[9] Because life as survival requires a susceptibility to death, because it requires openness to the other, because it requires, therefore, that life be precarious and fragile—well, for these reasons, to eliminate this susceptibility in immortality can only mean to die (or, more precisely, to be dead). This is a conclusion that, as we shall see, is no less necessary from a strictly biological perspective, even if its implications are not exactly the same.

One link between the biological and phenomenological approaches to life lies in the importance of contingency, or mutability, for each. The selfsame immutability of Descartes's substance is no more acceptable from a strictly biological perspective than it is from a phenomenological one. From the biological perspective, life is unthinkable outside of the metabolic activity that sustains it. And because metabolism consists of a constant exchange of matter and energy between an organism and its environment, immutability is equally unthinkable for it. An immutable, materially self-identical, or immortal (in the sense of absolutely unkillable) life-form is therefore a contradiction in terms. Nevertheless, even if the view that life is metabolism entails death's remaining a possibility for the living (in other words, even if this view also entails a certain a priori exclusion of immortality in its substance-materialist conception), the sense in which death exists as a necessary possibility for biological life is not simply equivalent to the sense in which death exists as a necessary possibility for Dasein, or for the being that has a relation to its own being.

Though we will return to the biological perspective and its implications for thinking mortality, it is worth noting here that the relation between strict biological fragility—or what philosopher Hans Jonas calls needful freedom—and the evidently more robust phenomenological fragility (or finitude) championed by thinkers such as Heidegger, Hägglund, and Nancy remains highly contentious. While there are many who would like to efface this difference, there are many others who would like to uphold it as absolute. In short, on this difference lies the question of the specificity of humankind and its particular distance from the rest of the living world. Even if a certain death is common to all life, the death that arises for Dasein (says Heidegger) has no simple equivalent for any other being. We might even say that it is the particular privilege of Dasein to invent a death nowhere else realized in the worlds of the living.

Phenomenological investigation, with all the insight it brings, thus forces one sooner or later to confront the problem of human/nonhuman animal difference, and with it the privilege allotted to the former. How to account for the evident distance between human civilization and all other, nonhuman forms of organization without either succumbing to the most vulgar form of anthropocentrism or risking the reduction of all emanations of life to a single undifferentiated presence? At times even Nancy yields to the former tendency—notably when he grants to humankind unique access to death, and by virtue of this access, sole possession of culture and community. The "relation to death" begets, according to Nancy, an absolute break between the mortality of human culture (with its community) and the "universal adhesion" of cancer, marked, evidently, by the self-presence of endless growth:

> Culture in general—all human culture—opens up the relation to death, the relation opened by death, without which there would be no relation at all: there would be only a universal adhesion, a coherence and a coalescence, a coagulation of all (a putrefaction that would always be vivifying for new germinations). Without death there would only be contact, continuity, and contagion, a cancerous propagation of life that would consequently no longer be life—or rather, it would only be life and not *existence,* a life that would not at the same time be *anastasis.* Death opens relation, that is, the division *[partage]* of departure. Everyone is endlessly coming and going, incessantly. Even that which appears as the end reveals itself without end. But this revelation reveals nothing, above all not a transfiguration of the dead into the living.[10]

Thus, while both biological and phenomenological thought let us think the necessity of mortality in life (and thus the a priori impossibility of substance immortality, understood as insusceptibility to death or change), as this passage makes all too apparent, one cannot simply equate their uses of the term. For the biologist, mortality refers to a physiological susceptibility, whereas for the phenomenologist, mortality instead refers to a relation to death—to a certain representation, or connection, within life, to a state of nonbeing that in fact interrupts representation and opens existence as thrownness, anxiety, and impossible communion. Even if, in the final analysis, any physiological susceptibility worthy of the name entails or requires a certain relation to death as its very point of departure (a question I hold in suspense), we must nevertheless respect the disparate intentions of each of these conceptions within their fields.

Cancers, as we shall see, may indeed possess the capacity for growth without end. Nevertheless, they are not, strictly speaking, insusceptible to death; they may be killed, as many oncologists will happily tell you. They are, therefore, not immortal in the Cartesian sense of the term, but neither are they strictly mortal—at least not if Nancy is to be believed—in the phenomenological sense. They do not have access—or, at least, we do not at present believe that they have access—to their own deaths in the mode of being-toward-death. They do not live as dying. Yet, because under the right conditions they can grow without end, we call them biological immortals, thereby indicating their specific difference from the mere biological mortal.

In the chapters to come, we will have to ask to what extent human culture—which, according to Nancy, is distinguished by its relation to death, by its being opened by this relation—can (or must) be kept apart from mere living cultures (such as cancers) or what he calls propagation without existence. We must ask to what extent life and existence can (or must) be kept apart in this way while remaining mindful of the double risk that lurks on either side of this decision: on the one hand, that of simply equating all forms of life, and in this way violently reducing the excess, nonpresence, or radical negativity that a robust phenomenological thought of death gives, and on the other hand, that of simply severing humankind from all other forms of life, and in this way blinding ourselves to the ways in which all life-forms—cancers included—are toward their own deaths, even if many may not have felt representations of this, their future demise, let alone a relation to the impossibility of a representation of what, in death, can never be made present. With or without minds, with or without brains, the living may be toward death, and as such, it calls for a more careful analysis of concepts able to bridge the discursive closures of phenomenology and biology.

To summarize: while a certain susceptibility to death (or material mortality) would apply everywhere that there is biological life, it is not as yet clear that what is called phenomenological mortality (or the experience of death and dying, as a mode of existence) can be generalized in this way.[11] What is clear, however, is that for both modern phenomenology and modern biology, to live is to be mortal, whereas immortality (understood as insusceptibility to death) is equivalent to being dead. In the words of Blanchot: "Dead—immortal."[12]

Substance immortality thus implies the necessity of a given substance—living or dead, animate or inanimate, material or immaterial—to remain itself in spite of every vagary. It is the negation (im-) of mortality understood as destructibility, which is why it may be applied to worldly beings as well as to intraworldly things. One may refer, for example, to one's immortal fame or memory, just as one refers to the immortality of the self or soul, by which is meant that fame will never die—that fame will remain essentially one and the same, even if I do die and I do change.[13] Critically, however, from the perspective of Heideggerian ontology (or, to be precise, from the interpretation of this ontological thought as it is found in the work of Hägglund), once this form of immortality is applied to living or worldly beings, it no longer appears to be desirable. This is because to be indestructible in this way would mean that, on the contrary, and as we have already observed, one were already dead.

I differentiate this long-standing interpretation of substance immortality (as well as its post-Heideggerian reception, as being equivalent to death) from what I am calling immortality's ontological interpretation, whose explicit, technical sense emerges first from within Heideggerian discourse. Ontological immortality no longer entails the negation of destructibility (which, Heidegger might respond, always reduces being to presence). Instead, it entails the negation of ontological being, as negation (im-) of mortality, understood as being-toward-death, or finitude. To become immortal in this way is to become incapable of the experience of dying. For this reason, ontological immortality does not have any direct bearing on one's susceptibility either to biological death or to physical destruction, which no one actually experiences, but only on dying *(Sterben)*—that is, living as dying, living as finite—because one's mode of being is born out of a fundamental relation to temporality. As Heidegger explains in *Being and Time:*

> Dying is something that every Dasein itself must take upon itself at the time. By its very essence, death is in every case mine, in so far as it "is" at all. And indeed death signifies a peculiar possibility-of-Being in which the very Being of one's own Dasein is an issue. In dying, it is shown that mineness and existence are ontologically constitutive for death. Dying is not an event; it is a phenomenon to be understood existentially.[14]

> Death is a way of being, which Dasein adopts as soon as it is.[15]

> Let the term *"dying"* stand for that *way of Being* in which Dasein *is towards* its death.[16]

Technically speaking, then, ontological mortality is a matter of access to the experience of death as a possibility, as such. Ontological immortality, by contrast, is a matter of the insusceptibility (or nonaccess) to this experience of death (as dying).

Naturally, given everything that I have already said about the problematic nature of human/nonhuman animal difference, one may well wonder whether such a hard-and-fast distinction between ontological mortality and ontological immortality is possible—whether, in other words, the line of articulation between the human animal and all other animal and nonanimal life can be so stark as to beget beings who have a relation to their deaths and beings who simply do not.[17] Derrida, of course, has questioned the straightforwardness of this opposition—without, however, wishing merely to efface it. He wonders whether such a privileged form of access to death, as such, in its authentic bearing, is ever possible, even in the best of circumstances.[18] Is death, he asks in the second volume of *The Beast and the Sovereign* and in *The Animal That Therefore I Am,* ever truly available to Dasein? That is, is death as such, in the mode of authentic being-toward-death, rigorously speaking, a possibility? Can Dasein truly die? Can it truly exist as dying, in the mode of *Sterben?* And if not, what happens in turn to all the subsequent distinctions that are grounded on this one?[19]

Whether or not such a distinction can be maintained, its influence remains far-reaching. Because the focus of ontological immortality is not on actual destruction or destructibility (as a present event, or *vorkommendes Ereignis*), but instead on mortality (*Sterblichkeit* or *Endlichkeit*) as the experience of the possibility (or impossible possibility) of such destruction, the animal, though just as fragile and precarious as we are, may nevertheless be said not to be mortal—at least not in the same way as Dasein, or man. From the ontological perspective taken by Heidegger and Nancy, an animal (or other life-form) that has no access to the impending of death on its life would be immortal. Insofar as it would not have its death as a possibility (or, strictly speaking, as an impossible possibility), it would exist without death (insofar as it could still be said to exist at all), and it would do so even if, as has been shown for many nonhuman animals, it nevertheless had a certain concept of death as a present event that comes.[20]

Heidegger, to be clear, never outright says that the animal is immortal. He is nevertheless quite clear about what distinguishes the mortality of man from the poverty of the animal, and it is this distinction that makes such a conclusion not just possible but inevitable. The animal, he explains, "perishes" *(verendet),* but it does not "demise" *(ablebt),* and it certainly doesn't "die" *(sterbt).*[21] In other words, its vitality comes to an end *(verendet),* but insofar as it does not live as dying *(sterbt),* it does not experience this departure as anything more than a present event (it doesn't *ablebt,* in other words). The difference between ontological mortality and ontological immortality (as Heidegger's readers have interpreted it) thus grounds the now familiar, if problematic, hierarchical differences among beings that are simply in the world, such as stones and other inanimate objects that are ultimately "worldless" *(weltlos),* animals that may *have world* but are, in the final analysis, "world-poor" *(weltarm),* and man or Dasein, who not only has world, but is also "world-making" *(weltbildend).*[22] Through the determination of mortality *(Sterblichkeit)* through finitude *(Endlichkeit),* which, according to Heidegger, can only properly be said of man, one may thus arrive at the conclusion of the de facto immortality of animals, insofar as the animal, while biologically alive *(Nur-lebenden)* and capable of a certain behavior toward death, nevertheless cannot conceive of the possibility of not being there. The animal cannot conceive of the possibility of its impossibility, and this is what authentic death (and mortality) names. The animal is thus deemed to be insusceptible to the experience of death as such, which is to say as the possibility of its impossibility, in the mode of anticipation *(Vorlaufen)*.[23] As Derrida puts it, "The animal is a living creature that is only living, as it were an 'immortal' living thing."[24] The animal, Heidegger might say (and even in direct contradiction to contemporary zoological knowledge), cannot be anxious; it has no sensitivity to ontological difference, or to the difference between being and beings.

It is, Heidegger explains elsewhere, verbal language alone that makes such access possible. Perhaps it would be better to say that, for the same reason that Dasein "has" language, Dasein is mortal. Thus, like god, the animal cannot speak, and the link between each of these predicates (speech and death) is essential for Heidegger, just as it was for Kleist. Speech and death are, Heidegger explains on more than one occasion, codeterminative:

> Mortals are they who can experience death. The animal cannot do so. But the animal cannot speak either. The essential relation between death and language flashes up before us, but remains still unthought.[25]

> This shows that language in all this is not subject to logic, and that a certain inconsequentiality belongs rather to the essence of language and its meanings. In other words, language is something that belongs to the essence of man in his finitude. To imagine a god expressing himself in speech is utterly meaningless.[26]

Ontological immortality would thus characterize the essence of the animal for the Heidegger of *Being and Time, Fundamental Concepts,* and *On the Way to Language.* The animal's specificity as a living being without speech is to be ontologically immortal; to be susceptible to perishing *(Verenden),* or biological death, without being susceptible to dying as such *(Sterben),* which is to say, to being in a relation to dying as such. To live as an animal is to live immortally.

Just as with Plato and Diotima, it is not a question—yet—of problematizing the metaphysical assumptions lying behind such a violent foreclosure of difference and diversity. At this point, it is instead only a matter of drawing out the lines whereby are demarcated the historical privileges of mortality, death, and subjectivity. Moreover, if I have privileged Heidegger as the originator of ontological immortality, it is because he formalizes these structures in ways that make them readily recognizable and translatable in his wake. It is not, however, because he is the first to recognize or to profess them.

The immortality of the animal as a matter of its ontological numbness *(Benommenheit)* or constraint *(Gehemmtsein)* is famously and problematically formalized by Heidegger. Nevertheless, the ignorance of the animal to death, and the resulting immortality that follows logically from it, are thoughts common enough to be found in the work of both Rousseau and Keats, if not long before. In *Discourse on the Origin of Inequality,* Rousseau declares that "no animal, merely as such, will ever know what it is to die," for "the knowledge of death, and of its terrors, is one of the first acquisitions made by man, in consequence of his deviating from the animal state."[27] In "Ode to a Nightingale," Keats famously takes the idea of animal ignorance of death and explicitly associates it with a form of immortality, now represented through his "immortal Bird":

> O for a beaker full of the warm South,
> Full of the true, the blushful Hippocrene,
> With beaded bubbles winking at the brim,
> And purple-stained mouth;
> That I might drink, and leave the world unseen,
> And with thee fade away into the forest dim:
> Fade far away, dissolve, and quite forget
> What thou among the leaves hast never known,
> The weariness, the fever, and the fret
> Here, where men sit and hear each other groan;
> Where palsy shakes a few, sad, last gray hairs,
> Where youth grows pale, and spectre-thin, and dies;
> Where but to think is to be full of sorrow
> And leaden-eyed despairs,
> Where Beauty cannot keep her lustrous eyes,
> Or new Love pine at them beyond to-morrow. . . .
> Thou wast not born for death, immortal Bird!
> No hungry generations tread thee down.[28]

For Keats, the nightingale—in contrast to the poet of the ode, who in fact worries himself to death over death—lives in ignorance of this specter. The nightingale knows nothing of "the weariness, the fever, and the fret" of man, nor of his corresponding "palsy shakes" and "groans." Of death as a possible event that either has happened or can happen (above all to it), the nightingale evidently remains in the dark. If, then, we are to understand by these lines that the immortal bird's immortality is a consequence of its ignorance of past "hungry generations" that do not "tread thee down"—above all in the form of a reflection that, passing through these generations, turns on its own finitude—then what is here proffered is a prototypical example of ontological immortality. It is not an actual physical or physiological imperviousness to death or destruction that makes the bird "immortal." Instead, it is only the absence of its access to death's possibility (or impossible possibility) that lends Keats's bird its Kleistian grace.[29]

Traces of ontological immortality can therefore be found broadly. Nevertheless, it is only in the wake of Heidegger's writings that one has come to dismiss substance immortality—not simply on the grounds of its technical impracticability or impossibility but because it is now believed to be undesirable in itself. This is of course Hägglund's position, and the difference—according to him—between a run-of-the-mill atheism and a radical one. Even were immortality possible, argues Hägglund, it would still be undesirable, to the extent that it would comprise a state of plenitude essentially equivalent to death. Thus, whereas proof of the immortality of the soul remains the goal for Descartes—to say nothing of contemporary Christian theology, let alone Kant and Hegel—for Heidegger and his followers, Dasein's aim can be none other than holding an authentic relation to its "ownmost possibility, which is nonrelational, not to be outstripped, and certain."[30] An authentic relation, in other words, to the possibility of "no-longer-being-there" that death evidently names.[31] To hold such a relation would preclude both substance and ontological immortality. It would do so on the grounds that the latter can only be construed as a privative state of apparent, animal ignorance, while the former can only be construed as nonbeing, or death.

5

BIOLOGICAL IMMORTALITY

The influence of the above understanding—of substance immortality as being equivalent to death—can today be found broadly. It is a given not only of the work of post-Heideggerian philosophers such as Derrida, Blanchot, Nancy, Todd, Butler, and Agamben, but also of evolutionary and developmental biologists, as early as Lamarck and Darwin and as late as Jacob, Dawkins, and Varela. Nevertheless, there are yet other senses of "immortality" today in use that cannot be derived from either Cartesian or Heideggerian senses of the term.

Both substance and ontological immortality can be opposed to what in the life sciences is called biological immortality. Biological immortality does not involve either a negation of destructibility or finitude. Instead, it involves only the elimination of the perceived necessity of a future death, on the grounds of physical decay. What changes in biological immortality is thus only the (perceived) fatedness of what lives to die, on the grounds of the inherent perishability (or divisibility, as Descartes might say) of embodied, corporeal form. It is a negation (im-) of mortality understood as death sentence, which negation may begin as a mere comment on material embodiment, but which cannot but culminate in visible changes to the comportment of what lives toward its life. For this reason, it must be treated both as a biological structure and as a phenomenologically significant possibility of worlding.

With biological immortality—but let it be understood that until the investigations of Part III, my use of this term is only provisional, referring to its current usage in the life sciences—we enter another order of our investigation. Naturally, we will ask in what the relation between biological immortality and substance and ontological immortalities consists. However, in doing so, we now leave behind the privilege accorded to the human and instead begin to address these questions from the perspective of the organism, or of life, in general.

As I have already indicated, there is a clear structural difference between biological immortality and substance immortality; the first is by nature indefinite, while the latter must be infinite and absolute. The difficulty, however, is that no merely analytic account of the structures proper to the ideal forms of these visions of immortality can account for the complexity of life as it is lived. Thus, even once we have worked out the abstract, theoretical sense of each of these modes of immortality, the problem will still remain of determining the ways in which life-forms live—live as mortals, immortals, or indefinite survivors, and live with mortality, immortality, or indefinite life. If, as I have argued by way of Derrida, sovereign positing is less an effect of an already constituted subject than it is a structural feature of the trace that begets subjectivity, then the question to which we must respond is ultimately more complicated than what Heidegger, Arendt, and Nancy are willing to admit. We cannot simply ask how each form of immortality ought to differ from the others. We must also eventually inquire whether, and to what extent, a kernel of the Cartesian interpretation of immortality (as indivisibility) does not already inhere in what we call, and recognize to be, mere biological life—not as an objective material attribute but as a significative element of its world.

Thus, once we have worked out the relationship between each form of immortality, we will still have to ask whether, and how, the problem of prosthetic immortalization haunts the living, as a consequence of the minimal projections of autodeixis or of the bare acts of autopoietic homeostasis. Only in this way can we understand in what consists the novelty of contemporary forms of transhumanism and posthumanism, which claim to invent immortality anew, and then grasp their specificity as perversions or mere extensions of a process perhaps begun no later than the first self-replicating molecules.

Biological Immortality and the Contingency of Death

As I mentioned above, the principal difference between biological immortality and substance immortality is that the former is not unconditional. One who is biologically immortal not only can still die in the sense of *Sterben* (living as dying, proper to Dasein), but also can still die in the sense of *Ableben* ("demising," or passing away, of Dasein), or *Verenden* (perishing, as is said of nonhuman animals).[1] Thus, the only difference between biological mortality and immortality lies in that one who is biologically immortal need not necessarily demise or perish, even over an indefinitely long timeline. Yet for this reason it is debatable whether biological immortality should still be called by this name. Should one even (still) use this phrase? The status of the designation—whether in the case of biological immortality "immortality" is a mere misnomer, propagated by unsophisticated biologists and huckster transhumanists, or whether its application to this mediate form of biologically conditional survival speaks to a tension internal to the concept itself, a kind of *Zwischenphänomen*—will be one of our questions. In any event, unlike substance immortality, biological immortality is not unconditional. Unlike ontological immortality, it does not require ignorance of death, or closure to the possibility of death as a possibility (or an impossible possibility), as such.

Instead, biological immortality only has a sense insofar as it can be opposed to a biological mortality, the principal feature of which—it is commonly thought—is that what lives is fated to die (now in the sense of demise or perish), on a sufficiently long timeline. As Wordsworth puts it, in speaking of the sense of such a fatality in his first *Essay upon Epitaphs,* "to be born and to die are the two points in which all men feel themselves to be in absolute co-incidence."[2] Thus, the sense of biological immortality arises and gains intelligibility only insofar as one considers biological life to be in and of itself a death sentence—insofar as mortal life, as such, is understood to be essentially fatal.[3] It is from the perspective of the inherent fatality of biological life that it becomes possible to say (for example, as Silenus does) that it would have been better never to have been born at all, for to be born (so the thinking goes) is to be destined to die. In the words of Lamarck, "All natural productions . . . have, as the whole world knows, the faculty of feeding themselves, of developing, of reproducing, and are necessarily subject to death."[4]

The prospect of biological immortality thus draws into question the reality of the perceived inevitability of death for biological life—its *tou bion,* as Plato would say, or prescribed period. It forces one to ask in what this inevitability consists—whether it is or was in fact a reality, and, if so, a reality for whom and of what kind? Is the apparent necessity of the inevitability of death of a physical or metaphysical nature? Is it a physiological and biomedical exigency, or is it an a priori, logical determination that follows from the nature of the living substance itself? Is it an inevitability for the human alone, for all animals, or for life in general? Alternatively, is not what contemporary molecular biology teaches just the necessity of judging the so-called inevitability of death on a case-by-case basis, and thus without generalization?

The (alleged) inevitability of death, which can be considered both biologically and phenomenologically, is only ever as firm as its underlying sociocultural basis.[5] Thus, for the Greeks, such inevitability was understood through the dual notions of *moira* and *bios.* In the myth of Er, presented at the conclusion to Plato's *Republic,* Socrates explains that the role of the fates, Lachesis, Clotho, and Atropos, is to establish the duration of each man's life prior to its being lived. Later, in the *Timaeus,* the titular protagonist explains that the limitation of life-span is one that applies to all life in general:

> For no diseases which do not involve great danger ought to be irritated by drugging. For in its structure every disease resembles in some sort the nature of the living creature. For, in truth, the constitution of these creatures has prescribed periods of life for the species as a whole, and each individual creature likewise has a naturally predestined term of life, apart from the accidents due to necessity. For from the very beginning the triangles of each creature are constructed with a capacity for lasting until a certain time, beyond which no one could ever continue to live. (89b–c)[6]

As Heidegger observes in *Being and Time,* in contrast to such classical approaches to finitude, today the "necessity" of death is understood as being of "'only' *empirical*" nature.[7] That is, its truth is taken to lie in empirical observation and analysis alone. In this way, the modern determination of death's necessity would draw from a different source than that used either by Plato or Wordsworth—or, for that matter, Heidegger. Where for Plato death was implicated in broader metaphysical considerations, including the relationship between the gods and man, and where for Wordsworth the ideas of birth and death—and later immortality as well—were thought to be offered in an immediate, universally accessible human experience, for the contemporary empirical sciences, death would instead be an emergent property of the specific structures at play within the living organism. Only from the perspective of these structures, which remain governed by physical laws even as they beget biomolecular informational relays that are irreducible to them, would death have any specificity whatsoever.

Before turning to the possibility—or, to be precise, to the plural possibilities—of biological immortality as they are given by biomolecular and physiological conceptions of life and death, we must first ask in what the distinguishing traits of an "'*only*' empirical" certainty in death consist. Before the specific pathways toward cellular immortality were offered through microbiological analysis, already the approach to death as a merely empirical, biological event had opened the door for the discovery of these pathways. The general possibility of treating death as an empirical event subtends the specific forms of biological immortality as they are put forward today.

Ironically, one significant consequence of the merely empirical determination of the necessity of death is its susceptibility to questioning and doubt—that is, to bringing into question this very necessity rather than simply reaffirming it. This may appear counterintuitive; if anything, what one expected from "objective," positivistic inquiry was just the rejection of every trace of immortality, at least in its theological determination, as the unverifiable projection of an eternal afterlife. Yet it is completely in line with the nature of empirical truths generally that rational analysis may lead to their reversal just as easily as to their demonstration. Empirical truths are by their very nature open to questioning and doubt. The treatment of death as an empirical, medico-scientific problem, we could thus say, is also coextensive with a certain bringing into question of death's (merely empirical) necessity as a biological event. Indeed, the birth of such (empirical) questioning is co-originary with this (measured) form of doubt. Let us now examine one such birth.

Perfectibility of Man

Since the 1950s, research in molecular biology has contributed multiple and diverse pathways to imagining how biological life-forms might live immortally. Based in the analysis of the widest range of living beings, from the greatest diversity of environments, it has discovered previously unimagined (and unimaginable) structures while demonstrating their inner workings with striking precision and clarity. Extensive knowledge of the functioning of cells—for example, with the discovery of the Hayflick limit—as well as study of the diversity of life—such as the vastly populous *Lomatia tasmanica* and vastly long-lived bristlecone pine—have thus provided explicit accounts of mechanisms through which something like indefinite survival can exist and might even become imaginable for us.[8] Molecular biology thus upends any sense of a universally certain fatality of life, just as it upends any sense of a universally applicable notion of self.[9] Moreover, the mechanisms it has discovered—telomere lengthening, genetic cloning, and sirtuin stimulation—are themselves highly generative. Not only do they demonstrate the diversity of fashions through which life sustains itself, but in so doing, they defy ready-made notions of identity and individuality, all while providing fodder for reimagining how humankind might itself one day become immortal.

Thus, genetic engineering generally, and CRISPR/Cas9 gene-editing technology specifically, give us to think not only that cell structure itself might be hypothetically modifiable at the level of the gene, but also that humans might feasibly enact the most minute of changes to their genomes in practice. In this way, modern biology gives birth to a huge array of previously unimaginable speculations. Whatever the status of these speculations—whether they remain for the most part fictions, whether they are based in faulty or misapplied notions of individual identity, or whether they do indeed have solid foundations in biomechanical and physiological principles—they nevertheless in each case adhere to a certain biological rigor.

Allowing, then, for the wildly generative nature of molecular biological research, we must nevertheless not give it exclusive credit for bringing into question the necessity of death. Even before practices of genetic engineering and theories of epigenetic manipulation became technically conceivable possibilities, as soon as death itself was seen as an empirical event—and hence as a medicoscientific problem—it in fact became possible to question its necessity on these very grounds. Thus, before we understood the specific mechanisms through which one might extend the life of an individual cell, and before we understood the relevance of DNA for cell division, the determination of death as a consequence of strictly physiological effects already opened it to forms of physiological treatment.

Such an approach is what we begin to see at work in the eighteenth century, within the now well-known debate among Godwin, Condorcet, and Malthus concerning the perfectibility of man.[10] As Maureen McLane writes of this post-Enlightenment "immortality controversy," "the immortality topos becomes, in this series of treatises and essays [by Godwin, Condorcet, Malthus, and Hazlitt], an arena for several contests: political, logical, scientific, physiological."[11] In it, we see proponents of perfectibility (in Godwin and Condorcet) contend that something like an indefinite prolongation of human life on earth should be possible, on the grounds alone of man's inherent, limitless potential.[12] Such a projection of human immortality differs categorically from the earlier—and still at that time prevalent—Christian confidence in the immortality of the soul. For Godwin and Condorcet, indefinite human life is a physiologically attainable goal. Though immortality itself was an old theological ideal, the grounds on which these writers now propound it are the very scientific principles so often thought to demonstrate its (theological) impossibility. Thus, at stake in their writings is the nature of the transition from immortality to indefinite life (or theological to biological immortality), as well as the epistemological grounds through which this difference becomes legible.

Ultimately, what the post-Enlightenment immortality controversy turns on is the difficulty of extrapolating what is reasonable for human reason to project. Put otherwise, it hinges on the discrepancy between understandings of what it is possible (and rational) to conclude about human life on the basis of scientific knowledge alone. When an economist like Malthus reads the claims of Godwin and Condorcet, what strikes him as scandalous is not the idea of immortality in and of itself but only what he perceives to be their careless misapplication of reason in arguing for the likelihood of such an indefinite prolongation of human life on scientific grounds. Thus, spurred by their writings, he argues in *An Essay on the Principle of Population* that by moving from a mere possibility to a probability, Godwin and Condorcet threaten the basic, foundational principles on which the sciences rest. McLane synthesizes their debate thus: "Malthus objected to the prominence of the unprecedented in Godwin's and Condorcet's thought. He understood that their speculations appeared on the border between the possible and the probable. If moral philosophy and political economy dispensed with the probable—as did Godwin and Condorcet in their immortality conjectures—then human science as a form of knowledge or a space of inquiry was itself imperiled."[13] What is critical for Malthus is thus to uphold the space of knowledge by, precisely, limiting its speculations.

For our purposes, more important than the specific arguments forwarded by each of these thinkers is the nature of the debate that they were engaged in. In each case, the basis of argumentation—everyone agreed—ought to have lain in observable experience alone. Ultimately, then, the difference between the two schools resides less in the role to be played by science, reason, or observation than in what one could reasonably conclude on the basis of such modes of inquiry. Thus it is that Malthus, on these grounds and in his typically hyperbolic fashion, contends that the resurrection of the spirit after death in fact appears "more probable" than any future immortalization of man, on earth:

> The only difference, with regard to our own apprehensions, that is not in favour of the latter assertion is that the first miracle we have repeatedly seen, and the last miracle we have not seen. I admit the full weight of this prodigious difference, but surely no man can hesitate a moment in saying that, putting Revelation out of the question, the resurrection of a spiritual body from a natural body, which may be merely one among the many operations of nature which we cannot see, is an event indefinitely more probable than the immortality of man on earth, which is not only an event of which no symptoms or indications have yet appeared, but is a positive contradiction to one of the most constant of the laws of nature that has ever come within the observation of man.[14]

In a scathing rebuke, spiritual resurrection is deemed by Malthus to be more likely than corporeal immortalization, on the grounds of the "most constant of the laws of nature that has ever come within the observation of man." Nevertheless, it was also on these grounds—of the observed "laws of nature," or, to be specific, now on the lack thereof—that Hazlitt would in turn respond to Malthus, arguing against the perception of just such a "fixed limit":

> With respect to the general shortness of human life compared with what it was in the first ages of mankind, this fact seems rather against Mr. Malthus, for if there is no certain date, no settled period to human life, beyond which it cannot hold out, but that it has varied from a thousand to a hundred years, so far there is no reason why we should not tread back our steps, or even go beyond the point from which we set out. There is no fixed limit: the present length of human life is not evidently a law of nature. The mere naked fact of its never exceeding a certain length at present is just as decisive against its ever having been longer, as it is against its ever being longer in the future.[15]

Thus, despite Malthus's protests, assertions of the possible and even probable immortalization of humankind remain conceivable on the grounds alone of the absence of any evidence as to a fixed limit. They remain conceivable, even if not necessarily justified, on the basis of a lack of evidence to the contrary. But whether we side with Malthus or Hazlitt, the very possibility of their debate has its roots in their shared empirical conception of death. Just as soon as death is viewed as an empirical, physiological event, the assertion of its necessity can only be as strong as the evidence (or lack thereof) put forward for it.

Though the terms of the debate have since changed, the basic premise of the argument between Godwin and Malthus remains just as relevant today as it was then: the gap between possibility and probability, or between what can be gleaned from the presence versus absence of evidence, conditions all projections of the future immortalization of humankind, on earth, considered as a biological organism for whom death exists as an (only) empirical, physiological reality. Indeed, as anthropologist Sarah Franklin has keenly observed, analogous transformations have occurred over the last two decades within the field of biology itself as we have entered "the age of biological control."[16] In the wake of successful experiments in mammalian cloning and the revelation that cell differentiation may not be irreversible, biology is no longer seen as a matter of "barrier[s] or limit[s]" but instead as an "entirely unconditional" field.[17] It is possibility rather than impossibility, in other words, that now marks the empirical study of life.

If all that death amounts to is an intolerable set of conditions for the survival of the organism, then it follows that one may hope to live indefinitely, so long as one removes all the physical impediments to natural, biological survival. Death is not a necessity—considered from the biological or physiological perspectives—but only a highly probable eventuality that is made all the more likely by current environmental and genetic factors. While Godwin believed that all such an indefinite prolongation required was an improvement in man's oftentimes gloomy disposition, today we might argue that it will require slightly more invasive, but no less realizable, measures, be it to our DNA or to the activity of our epigenetic survival circuits.

In what follows, we will ask in what these changes are supposed to consist and where the ideas for them originate. As we shall see, once death is determined to be a strictly empirical event, the figure of immortality does not just survive but thrives within medicoscientific discourse. Medicoscientific discourse invents new pathways toward the immortalization of life—or at any event, even if these pathways in many cases have precedents in other domains, medicoscientific discourse reinvents such pathways as technically achievable. In other words, it reinvents them as calculable eventualities, and it does so with ramifications for the longest tradition of immortalization efforts. In the twentieth century, microbiology and molecular biology have thus become the greatest peddlers in fictions of immortality as they weigh in on the differences between fiction and reality when it comes to life.

Next, in chapter 6, I return to the work of Derrida and Heidegger to see how this biomedical questioning of death interacts with their thoughts on survival, mortality, and existence. I then take a final look at the alleged difference between the desire for immortality and that for mere survival from the perspective of mortalists like Hägglund and May before turning, in Part III, to the multiple pathways provided by modern biology for thinking about the possible immortalities of biological life-forms. Such pathways always require biological definitions of selfhood, but these definitions are rarely straightforward. Biological research provides both the basis for thinking the identity of the organism and the basis for questioning that identity. The inability to define a clear-cut biological self does not, however, simply undercut immortalization efforts. On the contrary, the always already differential nature of organic life is the very condition for them. As long as the organism is understood to be a self-producing organized totality—be it autopoietic or homeostatic in nature—the indefinite translation of that self/totality will remain as a possibility, even if an implausible and perhaps impossible one.

6

MORTALITY REVISITED

On Derrida's "Condemnation to Die" and "Condemnation to Death"

The determination of death as a biomedical event permits its necessity to be questioned on biomedical grounds. Today, one does so on the basis of the most rigorous medicoscientific research. It would be tempting, therefore, to differentiate between two kinds of approaches as they relate to death as an ontic phenomenon: those that appeal to prescientific methods, such as, evidently, was the case for so many alchemical immortality potions and papal blood transfusions, and those that, precisely, use scientific methods, such as therapeutic techniques that engage bodies as biomolecular organisms. If this difference is on the one hand an essential heuristic for interrogating and categorizing distinct historical approaches to human mortality and immortality, especially as they begin to diverge in the eighteenth century with the birth of biology as a field, nevertheless, on the other hand, the appeal to such a difference will leave one blind to a more fundamental problem, which cuts across this divide—namely that any appeal to the minimal possibility of a solution to death, be it biomedically grounded or not, must make recourse to an extrascientific notion of life.

What I am now calling extrascientific is not simply opposed to the scientific or the empirical. Nor is it a mere placeholder for what one day might become a fully scientific concept. It is a matter, instead, of what must remain irreducibly extrascientific within any effort to extend the survival of the living. Every effort at life extension must determine, which is to say decide on, the essential versus nonessential elements of what lives. In doing so, it projects this survival as a minimally constituted self or individual. Yet the decision on the essence of the living, while perhaps informed by empirical research, cannot derive its ultimate principles from such research. Though metabolism, consciousness, and spatiotemporal continuity may serve as grounds for deciding on the life of what lives, no purely theoretical or empirical scientific principle can justify the decision *between* each of these standards, which in fact serve to perform (rather than merely reflect) the "life" in each case taken aim at. In sum, there can be no meaningful concept of either mortality or immortality without a prior determination of the life of what lives, and both pre- and postbiological approaches to life extension participate irreducibly in the performance of their referents. It is this common denominator that stifles any simple effort to establish a hard-and-fast divide between them.

Thus, if I am reluctant to assert a clear break between pre- and postscientific approaches to life extension, it is because in both cases, one has recourse to irreducible fictions of life. These fictions are not uniform. Nevertheless, insofar as in both cases fictions of life subtend visions of possible immortalizations—through, for example, selection of what is and is not to be preserved—we cannot begin to differentiate between the two until the irreducibly fictive character of each has been fully exposed. The question should therefore not be how scientific accounts of biological immortality dispense with the illusions of religious visions of immortality. Instead, we should ask in what ways the fictions of biological discourse rewrite and even repurpose the fictions of ontotheology.

Having admitted this, what is critical is to recall that the determination of the mere empirical reality of death's necessity—and the subsequent, and inevitable, questioning of that necessity—bear an important specificity with broad effects today. One will have to inquire both as to the historical limits of this perception and as to its ramifications, above all for the lived experiences of life and death—of death in life and life in death—that come in its wake. As we have just observed in chapter 5, what is called biological immortality has both a distinct historical specificity and a philosophical generality that defies this historicization. It is a mediate state between mortality and immortality, considered in their Cartesian, dualist interpretation as destructibility (or inevitability of destruction) and indestructibility (or inevitability of perdurance). For this reason, the designation may be applied just as easily to that which is ontologically finite as to that which is ontologically infinite—both, that is, to the Dasein that lives as dying and to the animal that lives as immortal (a category that includes, according to Heidegger, not only the mammal, but also the plant, the virus, and the bacterium). The question remains, however, how biological immortality might affect the experience of finitude. We must ask, that is, whether and to what extent it represents a modification thereof. If biological immortality names the empirical reality of a life without set term (such that there may be no term), then indefinite life may serve to identify the ontological or experiential reality of such a mode of existence.

In his late writings, Derrida frequently refers to the feeling of fatedness, or of the inevitability of death, which is commonly presupposed by the subject, as being *condamné à mourir,* "condemned to die," which condition he distinguishes from the juridical state of being *condamné à mort,* "condemned to death"—as happens, for example, through the imposition of a death penalty. "I am, we are all here, man and woman alike, condemned to die, but the chances are slim that any of us here will ever be condemned to death," he writes.[1] Being *condamné à mourir* differs from the Heideggerian being-toward-death *(sein-zum-Tode)* insofar as the latter refers only to a possibility (or to an impossible possibility) to which one is always in relation insofar as one exists, while the former refers to a state, or to an awareness of a state, in which this sheer but irrevocable possibility of death is also felt to lead, necessarily and eventually, to a realization of death, or no-longer-being-there, as a present event that will overtake one. For one who is *condamné à mourir,* death is thus always a future perfect; it is expected *(erwartet),* not just anticipated *(Vorlaufen in die Möglichkeit),* to make use of Heidegger's technical distinction.[2] It is experienced not simply as possible but as a possibility that will have arrived, become factical, and come to fruition, given a long enough timeline. Being-toward-death, as opening of temporality and finitude, does not dictate that one actually suffers demise *(Ableben),* or that one's corporeal being actually perishes *(verendet).* It dictates only that one is as a relation to one's possible nonbeing (death); that one lives as dying *(Sterben);* and that something always remains outstanding and incomplete insofar as one *is.* The experience of the condemnation to die (or death) takes this general condition of being as such and transforms it into what it is commonly experienced as: a future present event.

Significantly, for Derrida, the principal marker of distinction between these two condemnations is that in the second, through the phantasm of a death penalty, some sovereign ventures to foreclose the contingency that marks life by setting a definite time of death, now known in advance. Whereas in the average, everyday condemnation to die, one is sure of one's demising—but not when this demise will come—in the phantasmatic condemnation to death, the time and date of that death become definitively certain:

> The calculating decision, by putting an end to life, seems, paradoxically, to put an end to finitude; it affirms its power over time; it masters the future; it protects against the irruption of the other. In any case, it *seems* to do that, I say; it only seems to do that, for this calculation, this mastery, this decidability, remain phantasms. It would no doubt be possible to show that this is even the origin of phantasm in general. And perhaps of what is called religion.[3]

The calculating decision of the death penalty attempts to put an end to finitude. That is, it attempts to foreclose the always underlying possibility of death that subtends existence and gives life its chance by turning that possibility into a calculated, present event. Under "natural" conditions, in contrast, though one commonly believes that one is fated eventually to die (that is, that one's life will come to a clear and decisive end, even if another life, or an afterlife, awaits beyond it), because the time of that demise nevertheless remains undetermined and indefinite, so too does one maintain a wholly different phenomenological relation to it. The condemnation to death attempts to foreclose the finitude that is opened by being-toward-death and that the condemnation to die still seamlessly integrates, according to Derrida:

> The point is that it belongs to life not necessarily to be immortal but to have a future, thus some life before it, some event to come only where death, the instant of death, is not calculable, is not the object of a calculable decision. Where the anticipation of my death becomes the anticipation of a calculable instant, there is no longer any future, there is thus no longer any event to come, nothing to come, no longer any other, even no more

> heart of the other, and so forth. . . . "My life," and especially my life insofar as it depends on the heart of the other, cannot affirm itself and affirm its preference except over against this, which is not so much death as calculation and decision, the calculable decidability of what puts an end to it. At bottom, I would say by way of perhaps an excessive shortcut, that what we rebel against when we rebel against the death penalty is not death, or even the fact of killing, of taking a life; it is against the calculating decision . . . and the supreme form of the paradox, its philosophical form, is that what is ended by the possibility of the death penalty is not the infinity of life or immortality, but on the contrary, the finitude of "my life." It is because my life is finite, "ended" in a certain sense, that I keep this relation to incalculability and undecidability as to the instant of my death. It is because my life is finite, "finished" in a certain sense, that I do not know, and that I neither can nor want to know, when I am going to die. Only a living being as finite being can have a future, can be exposed to a future, to an incalculable and undecidable future that s/he does not have at his/her disposal like a master that comes to him or to her from some other, from the heart of the other.[4]

What do we rebel against when we rebel against the death penalty—when we rebel against the technology of putting to death that transforms the indefinite certainty of death into a definite, dated certainty? According to Derrida, what we rebel against is the foreclosure of a future that only finitude can give. Finitude births the incalculability and undecidability of a future. The phantasm of a death penalty is the phantasm of an end of such finitude.

Understood in these terms, the possibility that is announced by the concrete, biotechnologically achievable possibility of biological immortality would amount to a third alternative, a supplement to this Derridean opposition. One may not be condemned, either to death or to die. Biological immortality (and the indefinite life that ought to follow from it) involves holding a different relation to vulgar time, though not a different relation to temporality (or finitude) as such. What changes when one becomes biologically immortal is that the certainty in death as an event that will itself come on a long-enough timeline, this certainty becomes indefinite. Death is no longer a question of when, but if. One could therefore no longer speak of a "condemnation," as authors such as Shelley, Houellebecq, and Ishiguro have explored.[5]

In *Being and Time,* Heidegger takes up the alleged certainty in death that one commonly holds as Derrida's condemnation to die.[6] For Heidegger, this certainty in an event that has not yet arrived but that will eventually arrive becomes immediately translated into an experience of death's indefiniteness *(Unbestimmtheit).* The certainty of death's future presence (one's condemnation to die) thus becomes the very vehicle for the denial of this certainty, by way of the indefiniteness of its date. This is precisely how, Heidegger explains, the "they" succeeds in covering over the more primordial, more certain, and more angst-provoking, authentic experience of death as the possibility of my not-being-there—that is, as the possibility of one's impossibility:

> One says, "Death certainly comes, but not right away." With this 'but . . .', the "they" denies that death is certain. 'Not right away' is not a purely negative assertion, but a way in which the "they" interprets itself. With this interpretation, the "they" refers itself to that which is proximally accessible to Dasein and amenable to its concern. Everydayness forces its way into the urgency of concern, and divests itself of the fetters of a weary 'inactive thinking about death'. Death is deferred to 'sometime later', and this is done by invoking the so-called 'general opinion'. Thus the "they" covers up what is peculiar in death's certainty—that it is possible at any moment. Along with the certainty of death goes the indefiniteness of its "when." Everyday Being-towards-death evades this indefiniteness by conferring definiteness upon it. But such a procedure cannot signify calculating when the demise is due to arrive. In the face of definiteness such as this, Dasein would sooner flee. Everyday concern makes definite for itself the indefiniteness of certain death by interposing before it those urgencies and possibilities which can be taken in at a glance, and which belong to the everyday matters that are closest to us.
>
> But when this indefiniteness has been covered up, the certainty has been covered up too. Thus death's ownmost character as a possibility gets veiled—a possibility which is certain

> and at the same time indefinite—that is to say, possible at any moment.[7]

Through the prospect of a certain event of death whose date is indefinite—if not indefinitely deferred—one manages to evade the more fundamental, ontological certainty of death as a possibility that always lurks, explains Heidegger. One thus manages to ignore ontological certainty by conferring definiteness on an ontic demise that is itself temporally uncertain. To return to Derrida for a moment, it would be within the modalities of the certitude of death (i.e., demise), as an event, that the distinction between his dueling condemnations must be understood. The sovereign phantasm of a definite time of death would announce both the greatest denial of the authentic ontological sense of death (which is as possibility) and a kind of perverse affirmation of it (as a certain possibility). Under normal circumstances of everydayness, the mere condemnation to die, which one accepts as being certain but indefinite, and which one then makes definite so as to evade it all the better, would allow for the mere covering over of this underlying ontological possibility.

The question thus arises how we might understand the specificity of what I am calling indefinite life, as the limit case of certain projected forms of radical life extension, were it in fact possible (which is to say technically accessible) for us. How ought one understand the implications of indefinite life—by way of some form of biological immortality, for example—within the modalities of being-toward-death? On the one hand, indefinite life would announce an augmentation of the uncertainty of death as an event, insofar as this event would now be projected as an if, and not simply as a when: one may perish or demise, but one may all the same not perish or demise, even over an indefinitely long timeline. This new uncertainty would model more closely the sheer possibility of death as *Sterben*—which is certain in its ontological necessity as a possibility but uncertain and indefinite as a present event—than did Derrida's condemnation to die, which was certain (as a present event) but indefinite (in its time of arrival). On the other hand, even as it may model more closely something of ontological death's sheer possibility, as we shall see, indefinite life nevertheless also betrays this death, insofar as it distracts from the necessity of the possibility of not-being-there, through the fantasy of a perpetually-possibly-being-there factically. In this way, indefinite life would present a new mode of evasion of death's authenticity, even if, ontologically speaking, Dasein may always already be indefinitely mortal, given that finitude dictates nothing about the necessity of eventual perishing *(Verenden)* or demising *(Ableben)*. That is, finitude dictates nothing concerning the traditional view of biological mortality as fatal. We may, from a strictly Heideggerian perspective, always already have been biological immortals or indefinitely mortal—not because we could live indefinitely long (in fact) but because, qua ontological mortals, nothing about an eventually present perishing of our biological selves is dictated. This state of constantly impending death without the felt necessity of its event is what I call indefinite life.

Biological immortality imposes itself on Dasein as a question of its factical mode of being. It emerges, within the general condition of finitude, or being-toward-death, as a possibility of the being that questions its own being: "In being towards its death, Dasein is dying factically and indeed constantly, as long as it has not yet come to its demise. When we say that Dasein is factically dying, we are saying at the same time that in its Being-towards-death Dasein has always decided itself in one way or another. Our everyday falling evasion *in the face of* death is an *inauthentic* Being-*towards*-death. But inauthenticity is based on the possibility of authenticity."[8] Biological immortality, or the indefinite life that may follow from it (and like Derrida's dueling condemnations), acts on this plane. Opened by the higher order of certainty that is ontological finitude, which gives existence *as* the constant threat of nonbeing, biological immortality announces a mere modification of the indefiniteness *(Unbestimmtheit)* of death as a certain, factical event, viewed as a "constantly occurring mishap or 'case of death,'" or as a "well-known event occurring within-the-world."[9] Instead of imposing unconditionality, absoluteness, or atemporality, as ontological and substance immortality are traditionally thought to do, biological immortality and indefinite life only remove the (perceived) necessity of death as a future, present event to come. In biological immortality, death (or end) might come, but it might not. Nevertheless, death remains as a possibility, even a necessary possibility. What changes is only that this necessary possibility ceases to be understood as a possibility that will, necessarily, realize itself in a presence, as one might say of a fruit that ripens *(vollendet)*.

Moreover, if death does come under such a condition of indefinite life, it is now understood to come either from the outside, caused by a contingent external condition, as an event or an accident that befalls, or as an each time potentially preventable, even if internally imposed or preconditioned, limit. One therefore remains fated (or condemned) to the susceptibility to death—something could always go wrong; the universe itself could suffer its eventual heat death—but one is no longer fated (or condemned) to die (in Derrida's sense of this term). One must live dying *(Sterben),* but one need not perish or demise (*Verenden* or *Ableben*), which is to say that one remains fated to having death's possibility continue to haunt one; one remains mortal, or "a-mortal," as Harari has referred to this condition, but one is no longer—if one ever was—constitutively fated to its eventual arrival,[10] and hence to one's actually not-being-there. Finitude is not thereby eliminated. Biological immortality and indefinite life coexist with finitude.

Thus, although biological immortality is a designation made by a worldly being in finitude, it applies to both world-creating and world-poor beings—to make use, still, of Heidegger's problematic categories. A chicken, a virus, or a cell is no less likely a candidate for biological immortality than a human. Indeed, both virus and cell are far more likely candidates, at least once considered from a strictly empirical, physiological perspective. Be that as it may, in each case, what is eliminated in biological immortality is only the specter (real or imagined) of the necessity of an impending date of death. Death still impends. It lurks at every turn and around every corner, whether one knows it or not. But its arrival is no longer assured, if it ever was.

7

UNDECIDABLE DESIRE FOR SURVIVAL

Harari, Bostrom, Hägglund

We can apply the enormous leverage provided by the acceleration of technology. A notable example is achieving radical life extension through "a bridge to a bridge to a bridge" (applying today's knowledge as a bridge to biotechnology, which in turn will bridge us to the era of nanotechnology). This offers a way to live indefinitely now, even though we don't yet have all the knowledge necessary for radical life extension. In other words we don't have to solve every problem today. We can anticipate the capability of technologies that are coming—in five years or ten years or twenty—and work these into our plans. That is how I design my own technology projects, and we can do the same with the large problems facing society and with our own lives.

—Ray Kurzweil, *The Singularity Is Near*

If Kurzweil and de Grey are right, there may already be some immortals walking next to you on the street—at least if you happen to be walking down Wall Street or Fifth Avenue. In truth they will actually be a-mortal, rather than immortal. Unlike God, future superhumans could still die in some war or accident, and nothing could bring them back from the netherworld. However, unlike us mortals, their life would have no expiry date. So long as no bomb shreds them to pieces or no truck runs them over, they could go on living indefinitely.

—Yuval Harari, *Homo Deus*

Before turning our attention to the biological mechanisms of life and possibilities for extension that follow from them, I want to briefly return to a problem that has occupied us since Part I: the possibility of identifying a hard-and-fast divide between the desire for mortal and immortal life. Both existential analytics and the contemporary life sciences dictate that substance immortality is impossible. There can be no question of an immutable, indivisible presence when it comes to the living—or, ontologically considered, the dying. And while one may indeed inquire about the technical feasibility—or, more broadly, possibility—of something like an indefinite extension of physiologically grounded human life, our question now concerns neither the strictures imposed by basic human anatomy nor the limit between human and nonhuman beings. Instead, our concern is only to better understand what is called the desire for immortality and to situate this desire with respect to that for mere mortal life. How, in sum, ought one understand the difference between what has been called the religious yearning for immortality and the secular-atheist, or radical-atheist, desire for more mere mortal survival? Can the one be separated from the other, as tech futurists and radical atheists alike demand?

I have tried to show that from the viewpoints of Plato and Derrida, the answer must be no. For Plato, the desire for immortality (understood as absolute, indivisible self-presence) is the very condition of the desire for survival (understood as conditional, divisible, and symbolic participation in the divine). For Derrida, it is precisely the absence of this indivisible kernel of presence that is responsible for birthing what we call desire, which is always a desire for presence. How, then, do proponents of the mortalist line of argumentation distinguish them? That is, how do they distinguish not simply immortality from mortality (be it definite or indefinite in nature) but also the desire for the one versus the other?

For Hägglund, May and many others, it is essential that we favor attributes of fragility, mortality, vulnerability, divisibility, and fallibility, insofar as these are essential elements of who—and what—we are. Life itself is not thinkable without an inveterate divisibility (and hence destructibility). To this inalienable condition shared by all life, humanity then adds the additional knowledge of this ineliminable fragility, which it appears to possess uniquely as its own. It is essential, these thinkers argue, that we embrace this knowledge—knowledge of our mortality—and in so doing reject the very idea of immortality, above all in its Cartesian configuration, not only as a lived possibility (which, we may readily enough agree, is a lived impossibility) but also as a goal. Immortality is impossible—and, strictly speaking, undesirable, because any desiring being would desire itself out of existence were it to realize this end—and so, they contend, one should instead embrace fragility; one should embrace mortality; one should live one's being-toward-death authentically. If immortality is impossible (antithetical as it is to the very nature of life), then, these philosophers argue, it follows that it should also be discarded as an idea, an ideal, and an end. As if one could do such a thing!

Our question for this philosophical tradition is this: what can be hoped vis-à-vis the projection of such a fragile life? Can one truly wish to be mortal, and, if so, in what consists the difference between this wish and that for immortality? Conversely, is there something endlessly tempting about immortality that remains even in the wake of its impossibility, even in the full knowledge of its status as a bad object? If so, in what lies the root of this pull? Is it simply psychological, or does it somehow adhere to the basic structure of what we call life? Though we have established clear lines of demarcation between the formal structures of immortal, mortal, and indefinite life, at stake in this further question is the meaning of each for the being that lives and is concerned with its own being.

Let us put the matter succinctly. In the wake of Heidegger's existential analytic of Dasein, it seems as though the desire for immortality can only be equivalent to a desire for death. It is the expression of a death drive. What can be hoped for is, at best, more of this life, only ever indefinitely more, an indefinite deferral of this death that is always mine and that cannot be overcome, even if dying may itself be extended indefinitely. This is what both Heidegger and modern biology teach, each in its own fashion, and on this knowledge seems to be based the very difference between religious and secular approaches to these questions. We are thus left with a choice: either we admit that immortality and survival represent two antithetical ends *as* ends, which we could choose between, or else we must attempt to identify a continuity or undecidability between the one and the other sufficient to reveal the false nature of this choice.

Dataism, Radical Atheism, and Transhumanism

The writings of Nick Bostrom and Martin Hägglund are exemplary of two such trends among contemporary theorists who opt to oppose these ends. In the former case, it is through an appeal to modern empirical science that the futurist and transhumanist proponent of indefinite life does so. In the latter case, it is through the tenets of Heideggerian phenomenology and Derridean deconstruction that the philosopher emphasizes the essential role played by death for any conception of life. If I now consider these thinkers together, it is because, despite many differences in their work, in both cases, they reject immortality for the same reason: its alleged unconditionality. What we need, each affirms, is more life, but only conditionally. This is not, however, because we wish to die any sooner than is absolutely necessary—God forbid!—but because conditionality is just the law—the law both of finitude and the physical universe. It is this insight, then, above all others, that is supposed to separate their thinking from prior iterations of immortalist thought, especially as it is embodied in religious, ontotheological, and theological-political discourses.

For my purposes, what is critical is that the secularisms of both Hägglund and Bostrom hinge on the possibility of establishing a hard-and-fast divide between this new discourse of conditional, indefinite, mortal survival and an older discourse of unconditional, infinite, immortal life. They hinge, in other words, on the possibility of purifying the former's desire, expelling it of its unreason, and having it embrace a mortality no longer touched by the impossible, unrealizable, and in reality undesirable phantasm of immortality.

It is obvious that if we are to separate the desire for a metaphysical immortality from the desire for a rational mortality, then we must know in what exactly the former and latter consist. It is not sufficient simply to declare oneself for one or the other, because, as we have already seen, what is called immortality is itself a complex; it is an at once culturally determined and quasitranscendental (or structural) phenomenon that may take on any number of forms, depending on context. Thus, if we are to oppose the desire for one to the desire for the other, it must be because phenomenological (or empirical scientific) discourse can be excised of this phantasm of ontotheology, which of course requires that we first be able to identify in what the kernel of this desire lies. Having identified the essence of each desire, we may then purify ourselves of the undesirable one before finally modifying our behavior accordingly.

How, thus, to understand the relationship between the (ancient, religious, impossible) desire for infinite, unconditional, immortal life and the (modern, secular, authentic) desire merely for the most finite, conditional, mortal life? Without simply effacing the gap between these two positions, we must ask whether they truly constitute two. Could one separate the desire for the most life possible with the ineliminable but absolutely minimized possibility of death (that is, the desire to retain the perpetual possibility of dying without actually dying) from that for life without death, from that for deathless immortal life, from that for an absolute life that must, we are told, be in some way equivalent to death?

According to common wisdom—for example, that of the acclaimed author of *Sapiens* and *21 Lessons for the 21st Century,* Yuval Noah Harari—the answer appears to be yes. Yes, the desire for immortality and that for indefinite life can be opposed. In an interview given after the publication of his enormously successful *Homo Deus,* Harari expresses the situation succinctly, capturing what has become a commonplace approach in thinking through their relation. It is a matter, for Harari, of an historical progression. It is one that results from certain effects of scientific thought and that contributes to the fading of one set of beliefs and the rise of another. To the question "Wouldn't we live happier lives if we got over ourselves and accepted that we're all going to die?," Harari replies:

> Well, this was the common wisdom throughout history. And most people accepted it because they had no other choice. But people still had fantasies about immortality—they simply postponed it to the afterlife. Most people throughout history thought they would live forever, but not in this body or on this plane of existence. Rather, after they die, [they believed] they would reincarnate or go to heaven, or some such story which still promised them immortality. What's happening now is that scientific wisdom tells people all this was pure fantasy, mythology, and that there's no evidence that there is any existence beyond death. But at the same time, suddenly, the idea of

> extending our lives indefinitely is becoming more achievable, more practical. Not within the next 20, 30, 40 years, but definitely within a century or two.[1]

For most of time, for most people, Harari tells us, some version of a belief in immortality burned bright. This fantastic, mythological fantasy concerning the afterlife was, however, upended when "scientific wisdom" finally called it out for what it was and pointed out that there is simply no evidence for a posthumous existence. Thus, it happens that "now," today, "at the same time, suddenly," another idea has taken hold, a new one powerful enough to replace the old one. Coincident with the decline of a belief in an afterlife—but is it mere coincidence?—and (again) as a result of scientific wisdom, another idea, one that is not only "more achievable, more practical" but also more scientifically verifiable, has emerged. Within a century or two, Harari predicts—though we ourselves, we who live today, at the time of the publication of his book, will in all probability not be there to see it—we (the greater "we," the "we" of humanity and posthumanity) will nevertheless be able to extend our lives indefinitely. To live on indefinitely is not to live on immortally. There is a difference, grounded in the very difference between science and religion, that supports this division. Thus, because the two are in fact separated by the same abyss that separates knowledge and nonknowledge, our belief in the one is not equivalent to our belief in the other. (Let us not forget, however, that it is still very much a matter of belief. Perhaps a difference between faith and belief separates one from the other. But even if it does, we must not forget that no one has yet seen with their own eyes a biologically indefinite human life—unless, that is, human life has in fact always been indefinite. But in that case, where would lie the novelty?) Thus, even if biologists readily refer to such an indefinite life as immortality, we—along with Harari—know well enough that they are not simply equivalent. Nevertheless, the invention of indefinite life seems to hold a critical historical relation to this earlier immortality—if not also a symbolic one, as Malthus was already prepared to point out. The one just seems to replace the other, as if by chance.[2]

Even if we could look past the historical coincidence of the decline of one and the rise of the other—a coincidence that, it should by now be clear, is not even that, for forms of indefinite life and immortality have always coincided, which means that the coincidence is not some kind of radical break or seismic shift between them but rather is only a new preference for one where there had previously (apparently) been a preference for the other—thus, even if we could look past this "at the same time, suddenly," it would remain to be seen whether this shift in preference, which is nothing less than an epistemological shift between two systems of thought (what we could call religion and science), is actually sufficient to change the nature of what humans desire. Harari does not ask this question. For him, it is sufficient to observe a difference in the grounds through which one state or another might be predicted, or excluded, as an empirical reality. Harari does not ask why humans want to live on in the first place, and thus on what grounds they might come to prefer indefinite life (or amortality) to immortality.

As one never tires of observing, the virtue of indefinite life is that it is, to all apparent evidence, scientifically feasible—or, at worst, scientifically conceivable, given our present lack of evidence to the contrary (again: Malthus or Hazlitt). It is an engineering problem rather than a theological matter, as Harari explains:

> True, we still don't have the acumen to achieve this, but there seems to be no insurmountable technical barrier preventing us from producing superhumans. The main obstacles are the ethical and political objections that have slowed down research on humans. And no matter how convincing the ethical arguments may be, it is hard to see how they can hold back the next step for long, especially if what is at stake is the possibility of prolonging human life indefinitely, conquering incurable diseases, and upgrading our cognitive and emotional abilities.[3]

We do not, as yet, possess the "acumen" to achieve indefinite human life. Unlike, however, the immortality of the afterlife, promised in various ways by Judaism, Christianity, and Islam, "there seems to be no insurmountable technical barrier preventing us from producing superhumans" who would be capable of this novel form of existence.

Similar to Harari and his call for superhumans, Bostrom appeals to the benefits of a future superintelligence, and like Harari, he too places immortality in opposition to indefinite life:

> It is hard to think of any problem that a superintelligence could not either solve or at least help us solve. Disease, poverty, environmental destruction, unnecessary sufferings of all kinds: these are things that a superintelligence equipped with advanced nanotechnology would be capable of eliminating. Additionally, a superintelligence could give us indefinite lifespan, either by stopping and reversing the aging process through the use of nanomedicine, or by offering us the option to upload ourselves.[4]

As far as Bostrom is concerned, indefinite life-span is not immortality. Although any number of self-identified transhumanists do in fact confuse the two, Bostrom is too careful to risk such a conflation. He understands not only that each term bears its own distinct history but also that, philosophically speaking, there are significant differences between them. As his above remarks indicate, there are at least two pathways to attaining indefinite life-span. One may either eliminate the biological necessity of death through technology like nanomedicine, or one may upload our at-present organic-based lives into an inorganic substrate. Critically, although neither of these outcomes would necessarily lead to us living forever, they would allow us the possibility to do so—or, at any event, to survive until the heat death of the universe, which is certainly long enough.[5] Thus, in contrast to the unconditional immortality promised by the many mystified religions of the world, for which there is no road map save for that provided by faith and sacred texts, "indefinite lifespan" names for Bostrom a distinctly rational hope, attainable through strictly scientific means, and toward which humanity not only can but should work. Having rejected the immortality of theists, Bostrom, like many futurists today, champions an indefinite life, the historical antecedents of which are to be found in the notable Enlightenment thinkers whose work we have already briefly explored, and the possible attainment of which follows from the basic principles of already established medical and information science—or, at least, *might* follow from them. But the current state of this debate aside, what is philosophically significant in Bostrom's work is that indefinite, amortal, and open life refer to novel and noteworthy formations because they are, in principle at least, not technically impossible. In this respect, if in no other, they categorically differ from an immortality that has always presented itself as being unconditional, and therefore as antithetical to the very structure of the physical universe as we presently understand it.

In contrast to Bostrom, Hägglund has no such confidence in the unrelenting progress of technoscience. For him, as for a number of prominent contemporary philosophers, including Bernard Williams, Martha Nussbaum, and Todd May, what is critical is to reject immortality—not, however, because we cannot realistically hope to live forever on purely physical grounds but because the very goal of immortality betrays a fundamental misunderstanding of what is desirable.[6] Hägglund's argument, which he first puts forward in *Radical Atheism* before reformulating it in *Dying for Time* and then again in *This Life,* goes as follows. If by immortality we understand a state of absolute immunity and imperviousness to harm or death, then to achieve this state would be to contradict the basic structure of existence. It would be to render life meaningless because it would remove all risk, and thus remove any sense of project and any sense of future from the structure of our lives. As he puts it,

> These apparently contradictory claims become consistent if we apply my distinction between immortality and survival. Insofar as salvation is understood as the absolute immunity of immortality, it is out of the question. There can be no salvation, since nothing can happen without the greeting of an other that can come to compromise any immunity. However, insofar as salvation is understood as a survival that saves one from death by giving one more time to live, it is not out of the question. It is rather a precarious possibility that always can "be refused, threatened, forbidden, lost, gone" because of the infinite finitude of time.[7]

As such, to desire immortality is oxymoronic, for it is to desire a state without desire, and one—it follows—that would be akin to death. The dream of immortality must therefore be rejected on logical grounds, because even if we think that we desire immortality, what we in fact desire is to continue to survive as mortals.[8] This is why, as he puts it, we must now embrace a "radical atheism" in place of the traditional atheism peddled in past ages: "Atheism has traditionally limited itself to denying the existence of God and immortality, without questioning the desire for God and immortality. Thus, in traditional atheism mortal being is still conceived as a lack of being that we desire to transcend. In contrast, by developing the logic of radical atheism, I argue that the so-called desire for immortality dissimulates a desire for survival that precedes it and contradicts it from within."[9] Hägglund, like Bostrom, believes that the basis of his discourse exempts it from the misunderstandings that mark theological writings. Both support forms of life that are distinctly not immortal; in the process, they believe they provide many paths for evading the pitfalls suffered by immortality's classical as well as modern proponents. One must reject the false promise of immortal life, first because it is undesirable in and of itself (being equivalent to death) and second because it is unrealistic in principle (being impossible to achieve, on the grounds provided by empirical science, Heideggerian phenomenology, and Derridean deconstruction). What, then, ought one desire instead? Harari, Bostrom, and Hägglund all agree on this: more mortal life! Thus, despite their divergent points of departure, each ultimately comes out affirming essentially the same extensionist vision of life:

> Secular faith is committed to persons and projects that may be lost: to make them *live on* for the future. Far from being resigned to death, a secular faith seeks to postpone death and improve the conditions of life. As we will see, living on should not be conflated with eternity. The commitment to living on does not express an aspiration to live forever but *to live longer* and *to live better;* not to overcome death but to extend the duration and improve the quality of a form of life.[10]

> Death is not one but a multitude of assassins. Do you not see them? They are coming at you from every angle. Take aim at the causes of early death—infection, violence, malnutrition, heart failure, cancer. Train your biggest gun on aging, and fire. You must seize control of the biochemical processes in your body in order to vanquish, by and by, illness and senescence. In time, you will discover ways to move your mind to more durable media. Then continue to improve the system, so that the risks

> of death and disease keep receding. Any death prior to the heat death of the universe is premature if your life is good.[11]

The question thus becomes, in what consists the desire for more (indefinite) mortal life? In what consists the desire for extension? What is extension?

As I have already tried to show, the single decisive difference between what is called immortality and run-of-the-mill mortality appears to be the unconditionality of the former versus the conditionality of the latter. Immortality is interpreted as death because it is absolute, whereas mortality is interpreted as survival because it is conditional. As Hobbes, Descartes, and Derrida alike put it, the former ought to be marked by indivisibility and the latter divisibility. If one is to oppose the desire for the one with that for the other, it is therefore because one believes that, qua objects of desire, immortality has always named death (indivisibility), whereas mortality, or survival, has always named life (divisibility).

Even if this is the case, however, one will nevertheless still have a hard time simply equating Descartes's (evidently contradictory) desire with a strictly suicidal tendency. How could one reduce the expressions of those who espouse the desire for immortality to a mere desire for self-annihilation (assuming, of course, that such a thing is possible)?[12] The sheer specter of the perdurance of the undying, immortal substance—of a (present) remainder that remains—shares something with both the desire to survive (contingently) and the desire for death (absolutely). Descartes seems to think that it would be possible to perdure infinitely. Infinite perdurance is his goal (though he is certainly not alone in this). Thus, if one were to claim that, in spite of himself, what Descartes truly desires is just more mortal survival (and this, in spite of every testamentary evidence), then would not this interpretation also miss an essential truth of his discourse, and even *the* truth to which we must yet reckon? Extension and perdurance traverse both immortalist and mortalist visions of survival. Until we have grasped their value, we will have understood nothing of this problematic.

Thus, if I have opted to problematize this opposition, it is for the following reason: though immortality may name death, and though (indefinite) survival or (definite) mortality may name life, to desire one or the other is nevertheless never quite as simple targeting the first or the second. The desire for immortality appears to be that for infinite, unconditional perdurance. Yet the desire for survival (or more mortal life) seems to be the desire for finite, conditional perdurance. But when one desires conditional, finite life—when, that is, one desires more life—what is it that one desires? One does not desire death. Instead, one desires the most (mortal) life possible. One desires one's life not to end, although with the ever-critical caveat that it might. One does not desire that one's life ends but instead that this life continue, always and ever on the precipice of death, but without actually dying. There is a repression of death in both instances. If we have learned to wish for more (mortal) life today, and if such a life appears to be grounded in the firmest biological and phenomenological evidence, then can it truly be said that the desire of this desire is itself other than that of Plato, Augustine, Descartes, or Kant? The basis of our knowledge has shifted, certainly. But has that of our desire?

The conclusion to be reached from this is not that all desire is only ever for immortality, as Plato–Diotima–Socrates were ready to admit. It is instead that the semantic kernel of immortality—perdurance, persistence, identity, indivisibility, extension, contiguity, and so on—remains a driving force at play, perhaps in any self-organizing, self-orienting, or self-reproducing system. This is why it is irreducible, even if one may in fact enact it in the most varied ways, and even if, given this or that context, what one calls immortality is never quite the same.

To sum up: can one truly affirm mortal life? Can one express a desire for more of it without betraying a trace of the metaphysical lure that has haunted discourses of immortality for thousands of years? The appearance, fantasy, or simulacrum of an autoaffective, death-resistant ipseity, whose life would be extendable in principle, if not in fact, indefinitely if not infinitely, harbors this trace to the extent that it already announces itself as alive, as extendable, as, in sum, a party interested in being preserved. Yet even if we were to overlook this minimal positing of the sovereign self, in the final analysis, is not the difference between the hope, desire, wish, or preference to live as long as possible—but always possibly to die—and that to live forever—without the possibility of death, infinitely thin? Is not the desire of each—that is, *extension*—the same, even if its expression is not? Which is to say, does not the fantasy of the indefinite preservation of risk ultimately come down to the same thing as that of the infinite preservation of the self, so soon as one admits that the value of "indefinite" here is, precisely, to name the possibility of nonclosure, now on the condition of possible closure?

On the Religious of Religion

To conclude, I would like to briefly recall Derrida's discussion of religion in the first volume of *The Death Penalty*. If, as he there gives us to think, the tendency to immortality rests fundamentally in the structure of the trace, then we could no longer say that the immortality of the soul, or of God, is a primarily theological, or even human, proposition. Before there can be theology as such, it is just the originary structure of survival, or what we could think of as the very *différance* of life, which begets theology or religion, as an irreducible, always possible, autoimmunitary response to the nonpresence of the origin. Religion, or "the religious of religion," as Derrida puts it, would then be just one name for one of life's responses to this inexorable problem. It is that response whereby life comes to prefer something above itself:

> The example par excellence, in truth the essence of this force of "opinion strong enough to cause someone to espouse it at the cost of his or her life," of this sworn faith or this belief, is what is called religion, the religious, religiosity. And Montaigne is speaking thus of *all* religions, not only the Christian religion . . . every religion is capable, because it is the essence of religion; every religion is capable of preferring something else to life, at the cost of life. In other words, religion is or grants the surviving of survival. . . . The religious of religion is always the acceptance of sacrificial death and the death penalty, in the shadow of a sur-viving that supposedly is worth more than life.[13]

"Religion" henceforth names a response to the problem of survival. It is a surviving of survival, which grants a surviving of survival, but one whereby life now calculates a beyond of itself.[14] If this is what, for a dominant tradition of the West anyway, has been handed down as the specifically human dream of immortality—opposed not only to the mortalist embrace of finitude but also to the mere biological immortality of plant life, or the privative immortality of the animal, which does not have death—then Derrida's recentering of the religious will force us to readdress the nature of this so-called human distinction in at least two directions. On the one hand, we will have to attend to traces of religiosity possibly at work in all modes of (human and nonhuman) survival. On the other hand, we will have to reassess the nature of the difference between the dreams of immortality of human life and those of mere biological or nonhuman life-forms. Might not each share in these dreams, even if not as conscious projections? If the dream of an immortal afterlife can be traced back to the minimal projection of the self as a distinct (extendable) point or origin, then the prospect of immortality may be understood to be structured into the very tendency of life as trace—that is, in the inexorable movement of *différance.* There can, in sum, be no simple opposition between survival and immortality, because the latter is the expression of the former, and the former is always possibly a ruse of the latter, which means that we will have to rethink our most basic and fundamental understandings of life, finitude, and the infinite.

PART III

INDEFINITE LIFE IN THE BIOLOGICAL SCIENCES

> *Promises of science.* Modern science has as its goal the least pain and the longest life possible—that is, a kind of eternal happiness: to be sure, a very modest kind in comparison with the promises of religions.
>
> —Friedrich Nietzsche, *Human, All Too Human*

In explorations of Plato and Derrida, we saw why the division between immortal and indefinite life is a difficult one to sustain. I argued that on the basis of their work on generation and sovereignty, it cannot be a matter either of opposing or of collapsing these figures. Even if one could remove all the linguistic ambiguities that haunt their relation, the underlying structural ambivalences that constitute this field and give life to its obscurities would still limit any project of theoretical elucidation. According to a long line of philosophical and literary investigation, there can be no simple choice between linking and unlinking immortality and indefinite survival. Or, to be precise, there can be no simple choice between the phantasm of the one and that of the other, or between the sense of the former and that of the latter. Each lives off this relation.

Nevertheless, the identification of death's necessity with a biomedical condition—the determination of death as a merely empirical or physiological event—permits this necessity to be questioned on novel grounds. As we have already begun to explore, appeals to prescientific methods, such as alchemical immortality potions, blood transfusions, poetry, sexual reproduction, and ontotheological theories of the immortality of the soul, cannot simply be equated with biomedically sanctioned life-extending prostheses such as gene therapies, metabolism-slowing practices of intermittent fasting, sirtuin-stimulating supplements, and, even if they are not yet implementable, at least the ideas of cloning, brain uploading, and invasive practices of genetic engineering. Each of these latter techniques has the advantage that it approaches the self as an embodied organism (or as an emergent property of that embodiment) that must either be altered in its biomolecular configuration or else removed from its organic substrate altogether—that is, if its life is to be radically extended. One may reproach these techniques on any number of philosophical, scientific, and ethical grounds. One may view them with well-warranted skepticism. Nevertheless, they each have at least two advantages over their prescientific counterparts. First, the sophistication of their knowledge of the mechanisms by which organisms function far outstrips the former and supplies an altogether different epistemological basis for their claims. Second, on the basis of this knowledge, the radical extension of life in each case that is aspired to can at best be described as indefinite, whereas the former efforts all have as their ultimate aim infinite prolongations.

In Part III, we will explore in greater depth both the novelty and gravity of this biomedical insight into physiological mechanism. In particular, we will seek to answer the following set of questions. In what way do the insights of Plato and Derrida remain true for research in the biological sciences? Can these sciences sever notions of reproduction, life, and biological immortality from their antique metaphysical bases?[1] On what grounds can this research, in defining life, organism, self, and mortality, on the basis now of genetic, physiological, and thermodynamic principles, claim to liberate itself from the ontotheological immortalist tradition? But also, how are we to understand the relationship between biological insight into mortality, immortality, and indefinite life, and phenomenological insight into the same? Can these traditions be bridged? If (as we have just seen in chapter 7) our question was whether the desire for mortal and immortal life could be distinguished, in the chapters to come, we will now ask whether the biomechanical understanding of organic life, as (indefinitely extendable) process, can truly be distinguished from the ontotheological understanding of immortal life as infinite, absolute presence.

It will be a tenet of what follows that knowledge of biological mechanisms, especially as concerns empirically verifiable pathways of indefinite survival, represents important discoveries both scientifically and philosophically. They are today the clearest and most decisive avenues through which notions of the human, transhuman, and posthuman are undergoing processes of transformation, reinvention, and redefinition. They are today, moreover, the most obvious and compelling forms of what I term prosthetic immortalization, amid the ever-accelerating bidding war to secure human dominance over space and time. There is, therefore, a real imperative to think the implications of such newly possible—or at any rate newly imaginable—possibilities, of technically feasible forms of indefinite life, the potential of which, were they implementable in principle as well as in practice—and perhaps even only in principle—would have ramifications for the whole history of human life and existence as we know them.

That being said, the production of such knowledge within the sciences must ultimately be treated alongside—indeed, as part and parcel of the very same lineage and history as—prescientific and precritical notions of immortality, such as those of the soul, everlasting fame, and hereditary reproduction. To treat each of these traditions together in this way is to accept that there can be no strict nor simple separation between their discourses, be it on methodological or other grounds. It is to deny the oppositions proposed, and in large part taken for granted, by philosophers such as Bostrom, Hägglund, and Harari. It is not, however, simply to conflate religious confidence with scientific evidence. Nor is it to overvalue the merely symptomatic (mis)use of the term "immortality," which pervades each. It is instead to take seriously the role that each plays in constituting its object, be it life or the soul, as well as the common basis on which both do so.

First a quick word about this basis as it infiltrates both scientific and theological discourse. I am not arguing that the objects of biological science and religious discourse are simply the same. Instead, when it comes to discussions of life extension, in both cases, speculations depend on opposing that which lives to that which does not. By way of this opposition, each discourse institutes its own stable and a priori extendable object of interrogation. Thus, it is the opposition between living and nonliving that secures for each field its object as a form of presence (be it purely sensible or purely intelligible), and that in turn supplies a common basis for both, even if the determination of the essence and nature of the "living" in each case differs.[2]

Thus, it is insufficient to say that life just is chemistry, or that the living is reducible to the (nonliving) physical interactions of its component parts, in order to break free from the deeply entrenched, metaphysical opposition between life and death that undergirds both notions of the soul and vitalist approaches to organic life. On the contrary, it is today precisely by way of such a reduction, as well as the confidence in positive difference that comes along in its train, that the division between a materially verifiable life and a materially verifiable nonlife becomes authorized.[3] In short, abandoning the dogma of vitalism is not sufficient to free bioempirical notions of life from many of the problematic oppositions that marked their antecedents.

For our purposes, what is critical is that just as soon as the opposition between life and nonlife is accepted—be it on vitalist or biomaterialist grounds—it is but a short step to the extension of the living (present), in the first case infinitely and in the second indefinitely. What is decisive is not the modern resignation to the tenets of empirical science in physicalism. This resignation itself allows for its own set of delusive fantasies—for example, in the reduction of all biological organization to mere information. The decisive step instead lies in the presumption of—or perhaps I should say decision for—a clear opposition between the living and nonliving. It is this presumption that in turn authorizes every logic of extension, and it is this extension that underlies notions of both infinite and indefinite life, thus marking the true kernel of immortalist thought. Only if it were possible truly to problematize the presence of life; only through a thinking of life–death that risked the very life it sought to define; only by denying the simple opposition of life with nonlife—only then could we begin to speak of an other of immortalization.[4]

For modern biological thought, what is indivisible is thus no longer the soul (which had offered absolute protection for the essence of the living, requiring that it be extended infinitely, whether this infinity be interpreted inside or outside time) but now the life of the physiologically functioning organism itself (which physiological interpretation secures the integrity of the living, doing so over an indefinite extension, but only so long as that functioning persists). It is the observable functioning of the physiologically living organism that today supplies the basis for discriminating between that which is and that which is not alive.[5]

Both modern biological thinking and much-decried vitalist approaches to life thus share this oppositional tendency—a tendency that is symptomatic of something deeper than mere prejudice and that one could scarcely imagine escaping entirely. The difficulty with trying to escape the opposition—life and death, life and nonlife—is that it is not simply a product of either theoretical analysis or theological inclination. It is not merely the result of this or that cultural, philosophical, or epistemic stance. As a robust tradition of biologically inspired phenomenological work has shown, the opposition between life and death is not abstract but a direct product of the processes endemic to what we call the living. That is, it is a product of the basic structures that constitute organic form, without which there could be no phenomenon of life as we know it. The opposition between life and death, however deconstructible, unverifiable, and ultimately without presence, is not merely a superposition on the living but also an artifact of the very self-organization that we have come to recognize as being particular to living systems as such.[6]

This is what the work of philosophers, biologists, and neuroscientists like Hans Jonas, Humberto Maturana, Francisco Varela, Evan Thompson, and Antonio Damasio has variously shown. Phenomenological identity is not simply epiphenomenal to biological structure but one of its emergent properties. This is not to say that such identity is ever stable, or that all organisms partake of identity, or of self, in the same fashion; it is not even to say that where there is phenomenological identity, the organisms productive of it are simply reducible to it. Rather, it is to say that where there is identity, be it consciously lived or unconsciously enacted, the sources for it can be traced back to basic and—so far as we know—universal characteristics of life such as metabolism, homeostasis, and autopoiesis.[7] Cognition, we could say, is coemergent with life; it produces or performs the difference between self and nonself, just as it does that between life and death, inside and outside. Whether or not there is something called life—a question that, for essential reasons, we must hold suspended—what is evident is that what we call life is this effect of self-organized structure.[8]

Ultimately, then, what we must learn to do—and what is of the greatest difficulty to do—is balance this evidence and what it betrays of the life/death opposition (now along the lines of beings with world and things that are worldless) with a deconstructive skepticism that neither affirms this difference as absolute nor denies this difference altogether, such as one finds with various dualisms and positivist materialisms. This opposition, as we shall see, is infinitely deconstructible, but being infinitely deconstructible, it is also infinitely resistant to deconstruction. Prostheses of immortalization thrive on this difficulty. Their survival is guaranteed by the impossibility of its success. However, so long as we insist on denying the deep ties that bind physiological mechanism to ontotheology, we will remain helpless to identify the root of the problem.

I now turn to the specific mechanisms that modern biomolecular science elucidates and the possibilities of survival that it canalizes therethrough. In this way, I will both clarify the meaning of the various forms of biological immortality identified by biologists, and later, in chapter 13, I explore how biological approaches to selfhood may supply similar insights into the nature of ipseity as deconstructive approaches to sovereignty.

8

GENETIC IMMORTALITY

As first explored in chapter 5, unlike both substance and ontological immortality, biological immortality is always a question of life-span, and life-span, as Nietzsche and Heidegger remind us, is today above all a biological concept. Though we have just seen how the possibility of an indefinite prolongation of human life seems to follow conceptually just as soon as that life is determined physiologically, nevertheless, at the same time, the application of the figure of immortality to biological forms gains its specificity and rigor only with the discovery of the particular mechanisms by which organisms propagate and survive. We must therefore examine the units of this alleged survival. This means studying both how discrete beings live without death—to the extent that they do—and in what the integrity of these beings consists in the first place.

No exhaustive account of either the means of survival or the units of identity of organisms is possible. This is the case both for pragmatic reasons of time and space and for essential reasons, because the set of each is constitutively unlimited. In what follows, I will therefore select a few examples for reflection. These examples have been chosen either because they play outsize roles in the history of biological thought or because they are outliers, which hint at the total range of possibilities. Again, the purpose of furnishing these examples is less to weigh in on the possibility or feasibility (or impossibility and unfeasibility) of something like human immortalization. It is rather to see how the field of currently imaginable possibilities is itself formed for technoscience—on what basis, through which economies of life and death, and to what as-yet-unrealized ends. Only once this prior basis has been uncovered can the meaning of such a possibility be properly understood.

Immortality has always been, and will always be, a matter of belief. The key to understanding which beliefs are today credited lies in the biological conception of life.

Knowledge of molecular biology produces novel conceptions of life that in turn supply previously unimaginable translative possibilities for the living. Well short of the transhumanist fantasies of dataism and brain uploading—which Lyotard, Hayles, Thompson, and Oyama have each shown repeat the errors of Cartesian dualism, now in terms of ill-defined notions of information—empirical-physiological accounts of organic systems are already sufficient to upend notions of a preset, internal, or universal limit on life-span.[1] In this way, empirical-physiological accounts open the door to the (at least theoretical) possibility of indefinite, embodied, and biological survival. It is the specific novelty of the latter—both its generative nature and its ties with the tradition preceding it—that we must interrogate, first of all by asking what biological survival means.

Questions of immortality, just like those of life, death, and self, are always a matter of scale within biological thought. Certain properties only emerge as properties within specific organizational limits. This means that although there are indeed certain constants that pervade all known instances of life, life itself, as we know it, is itself only intelligible starting at a certain level of organization—that of chemical systems—and up to a certain level of organization—that of planetary systems.[2] Moreover, within these bounds, the basic units associated with living systems may be found in any number of orientations, from the relatively autonomous bacterium to the symbiotic lichen and the superorganismal insect colony. Moving from one level of organization, or scale, to another (for example, within a single organism), or from one mode of being to another (for example, between different types of organisms), requires identifying specific structures that may either be shared or unshared across these divides. Whether one could ever fully justify one's reliance on a given structure for the purposes of defining life is today a question of the highest importance, and one hotly debated among astrobiologists.[3] For our purposes, what is essential is not the theoretical possibility of defining life once and for all, but instead the precedents already in place for doing so, as well as their ramifications for medicoscientific speculations on life, survival, immortality, and death. Today, the most widely accepted signs of life are metabolism and reproduction. "Life," according to NASA's current working definition of the term, "is a self-sustaining chemical system capable of Darwinian evolution."[4] The overwhelming reliance on these two handles of vitality allows us to put forward a provisional typology of the forms of immortality that are rendered thinkable by modern biological thought. In thinking, or making thinkable, these forms, modern biological thought invents them as potentially realizable empirical possibilities.

Roughly put, we can distinguish between two general types of immortality that are operative within the contemporary life sciences: genetic and metabolic. I say "roughly put" because, as we shall see, neither of these forms can ever be entirely separated from the other. In practice, within the being of the beings that live, they are always intertwined, which means that each concept (gene, metabolism) is defined through, or with an eye to, the other.[5] Moreover, though the bases on which each concept is founded are, on the surface, at odds, it remains to be seen whether the same ideal does not ultimately ground both.

With that said, and with this fold of complexity admitted, it nevertheless remains the case that we must distinguish between two general forms of immortality that predominate in biological discourse. What I have called biological immortality will henceforth refer specifically to the metabolic variety, because only this one can be said to coincide with what we generally call living. We will break down this still overly broad category into the three specific kinds that compose it: unicellular, multicellular, and organismal. Genetic immortality, in contrast, refers to the survival of the individual gene, or by extension to that of the genome as a whole. (As we shall see, however, in principle, it could be extended to any informational unit capable of Darwinian evolution by natural selection, including memes of all kinds.) To use a distinction offered by scientists Sara Walker and Michael Lachmann, to which we will return, while genetic immortality partakes of life, it cannot, strictly speaking, be said to be alive.[6]

One can locate a precursor to the modern notion of genetic immortality in the work of German naturalist and evolutionary biologist August Weismann, whose 1885 *Continuity of the Germ-plasm as the Foundation of a Theory of Heredity* and 1892 study *The Germ Plasm: A Theory of Heredity* develop a theory of heredity through the hypothesis of a difference between germ cells, which transmit traits across generations, and differentiated, functional somatic cells. While for Weismann, somatic cells are mortal—or, at any rate, at least as mortal as their individual bearers—germ cells are "potentially immortal" because they, like their unicellular predecessors, can divide without end, and thus they live on in all future members of a given lineage.[7] What allows for heredity truly to function, however, is not the power of the germ line simply to propagate. Although the propagative power of the germ line—itself still an example of metabolic biological immortality—may be necessary in order for the reproductive machinery of organisms to persist, it is not, according to Weismann, in and of itself sufficient to account for the fidelity of ontogenetic development that persists across generations. Instead, this crucial role within heredity falls to the germplasm, which it is the function of the immortal germ cells to bear:

> The nature of heredity is based upon the transmission of nuclear substance with a specific molecular constitution. The substance is the specific nucleoplasm of the germ-cell, to which I have given the name of germ-plasm.[8]

The germplasm is neither a self-sustaining organism nor a fully functioning cell; rather, it is a nuclear substance contained within certain cells, which serves to direct development. The germplasm, more so even than the germ cell, is therefore responsible for the propagation of species because it is what dictates morphology and what, from within the cell, conditions similarities among members of a clade:

> I propose to call it the theory of "The Continuity of the Germ-plasm," for it is founded upon the idea that heredity is brought about by the transference from one generation to another, of a substance with a definite chemical, and above all, molecular constitution. I have called this substance "germ-plasm," and have assumed that it possesses a highly complex structure, conferring upon it the power of developing into a complex organism. I have attempted to explain heredity by supposing that in each ontogeny, a part of the specific germ-plasm contained in the parent egg-cell is not used up in the construction of the body of the offspring, but is reserved unchanged for the formation of the germ-cells of the following generation. . . .

> The hypothesis of the continuity of the germ-plasm gives an identical starting-point to each successive generation, and thus explains how it is that an identical product arises from all of them.[9]

The germplasm is thus that which is both heritable (in principle) and inherited (in fact) from one generation to the next. It is that which remains constant and the same across generations—determining, for example, phylogenetically stable morphologies—while the mortal soma comes and goes with the life and death of each individual. Germplasm is not alive in and of itself (in distinction to both germ and soma, it is a mere molecular component of the cell), but it is nevertheless an iterable presence whose heritable, phylogenetic stability ensures the ontogenetic invariabilities of individual members of a lineage.

While much of what Weismann believed concerning the strict division between germ and soma has since been called into question by subsequent experimental work, the idea of a nuclear substance whose identity across generations conditions like development today lives on, and it does so, above all, through the more refined, biomolecular notion of the gene.[10] Today, genetic immortality, subtended by "selfish" replicator genes, is the notorious hobbyhorse of evolutionary biologist Richard Dawkins. As Dawkins explains in the introduction to the anniversary edition of *The Selfish Gene*, reflecting, now with hindsight, on its polarizing title: "Given the dangers of that style of error, I can readily see how the title could be misunderstood, and this is one reason why I should perhaps have gone for *The Immortal Gene*."[11] Genes are immortal for Dawkins not because they always remain the same but, precisely, because they have a great potential to do so. Like Weismann's germplasm, the gene is what persists across generations, conditioning development and preserving identity despite difference. Although the gene's chemical basis, in deoxyribonucleic acid, which is to say DNA, shields it from the simple, ontotheological presumption of strict immutability, such as one finds in Descartes's immaterial substance, its immortality is nevertheless an immortality of stasis and stability. The gene does not persist as a single surviving material substance or substrate. Instead, it persists as an infinitely repeatable informational copy. The replicability of information stored in individual genes is what renders them immortal.

By translating Weismann's heritable germplasm into the biomolecular language of the genetic code, modern biological science is thus able to locate an immortal kernel at the heart of mortal life. It does so by locating something nonliving and potentially indivisible within the cell: information. In the words of Dawkins:

> Natural selection in its most general form means the differential survival of entities. Some entities live and others die but, in order for this selective death to have any impact on the world, an additional condition must be met. Each entity must exist in the form of lots of copies, and at least some of the entities must be *potentially* capable of surviving—in the form of copies—for a significant period of evolutionary time. Small genetic units have these properties: individuals, groups, and species do not. It was the great achievement of Gregor Mendel to show that hereditary units can be treated in practice as indivisible and independent particles. Nowadays we know that this is a little too simple. Even a cistron is occasionally divisible and any two genes on the same chromosome are not wholly independent. What I have done is to define a gene as a unit which, to a high degree, *approaches* the ideal of indivisible particulateness. A gene is not indivisible, but it is seldom divided. It is either definitely present or definitely absent in the body of any given individual. . . . Another aspect of the particulateness of the gene is that it does not grow senile; it is no more likely to die when it is a million years old than when it is only a hundred. It leaps from body to body down the generations, manipulating body after body in its own way and for its own ends, abandoning a succession of mortal bodies before they sink in senility and death. The genes are the immortals, or rather, they are defined as genetic entities that come close to deserving the title.[12]

The potential immortality of the gene is that of a particular informational structure able to be reproduced with exceptionally high fidelity through the reliable replication of chemical DNA sequences. Indeed, as Dawkins readily admits, there is no clear-cut biological standard by which individual genes can be identified. It is not as simple as saying that an individual gene codes for an individual protein, for example.

To speak of individual genes is instead to use a heuristic, one based on a given unit of DNA's causal role in selection. A gene, for Dawkins, is nothing but a hereditarily stable selection of DNA that will have survived because it positively affects its survival rate by propagating copies of itself through diverse and variegated phenotypic effects.

Like the text of a book or poem that lives on in each manuscript, typescript, and digital printing, the gene can live on, immortally, in each cellular copy. In this way, it is able to survive the deterioration of its physical substrate because, as David Haig clarifies, its existence as an "informational gene" is no longer bound to the specific fate of each of its individual "material" iterations.[13] Genes as texts, we could say, are potentially immortal, whereas individual, material copies of genes remain mortal and finite, insofar as their survival is bound up with the fates of their mortal bearers—or, as Dawkins likes to call them, their "survival machines."[14] As Daniel Dennett writes, expanding on the topic in *Consciousness Explained:* "Memes, like genes, are *potentially* immortal, but, like genes, they depend on the existence of a continuous chain of physical vehicles, persisting in the face of the Second Law of Thermodynamics. Books are relatively permanent, and inscriptions on monuments even more permanent, but unless these are under the protection of human conservators, they tend to dissolve in time. As with genes, immortality is more a matter of replication than of the longevity of individual vehicles."[15]

The notion of genetic immortality, in its biomolecular configuration, is thus inextricably bound to the replicability of information. Just as a meme may be reproduced, with fidelity, across space and time, so too is a gene nothing but the singularity (or particulateness) of its code, which therefore is no more bound to any one of its material embodiments than it is to any other. The privilege of information, at least since Claude E. Shannon's 1948 "Mathematical Theory of Communication," is tied to its fidelity across generations. Although the existence of the informational gene still depends on what Dennett calls the "continuous chain of physical vehicles" that perpetuate it, unlike Weismann's germplasm, which still remains fixed within the physical lineage of its descent in germ cells, the gene's immortality, as essentially informational, would no longer need to rely on any strict physical contiguity between its vehicles. In other words, as informational, the gene's replicability is conceived as essentially disembodied, even if it is always housed in one physical substrate or another, and even if it is always connected to some prior iteration through a chain of energetic informational relays.[16]

Because of the nature of the division operative in the working opposition between an (immortal) informational and a (mortal) material gene, this line of inquiry has often been accused of repeating a Cartesian-style dualism. Leaving aside, for the moment, questions about the ramifications of such an association, what should by now be clear is that the genetic immortality of Dawkins, Haig, and Dennett is, at the very least, modeled after Descartes's notion of immortal substance, which, as indivisible, immutable, and selfsame, attains immortality by being identical with itself. Nor is the fact that Dawkins and Dennett both admit, in the case of the gene, that it is only a potential indivisibility, or a potential immortality, sufficient to distance them from this characterization. It is precisely the ideal, implicit in this potential, of the absolute fidelity of a lineage with its origin that orients the notion of the gene for them and lends it its intelligibility.

Thus, even if Dawkins acknowledges that his is an impossible standard, and even if Haig clearly understands that the chemical code through which DNA is formed attains an at best differential identity, replication without difference nevertheless preorients the itinerary of both theorists and remains the safeguard of everything to come. Why, after all, even appeal to the figure of an immortal gene unless atomistic self-identity is the standard by which it is to be measured? Whether or not, then, particular genes ever live up to this standard, conceptually speaking, their informational specters are defined through its ideals of presence and saturation.

9

BIOLOGICAL IMMORTALITY I

Unicellular (Metabolic)

> At any rate we only know of potential immortality in the cells which constitute independent unicellular organisms, and the nature of these is such that they are continually undergoing a complete process of reformation.
>
> —August Weismann, "The Duration of Life"

Genetic immortality consists in the fidelity of an essentially informational structure that can be reproduced—or replicated—with potentially little or no change, over an indefinite number of generations. Its immortality is an identity of form, not matter; it is a qualitative, but not numerical, identity, to adopt the language of Parfit.[1] In this way, the gene is thought to be like a text, poem, or meme: an infinitely iterable segment of data.

Although, then, the gene that is selfish is not motivated and does not act alone (it requires an elaborate, multiorganismal, and often multispecies interaction to bring about self-replication), its very legibility as a gene nevertheless requires reference to its causal role within this iterative process. Thus, in distinction to the wholly inanimate (i.e., nonreplicating) electron, the inertness of the gene's code is in some sense enlivened through its role in self-replication. By way of this role, it becomes legible as a constructor, in David Deutsch's sense of the term.[2] This also makes it, in Stuart Kauffman's parlance, part of a Kantian whole.[3] The gene is both for and by means of itself. Its being—the fact that it *is*—depends on its prior, functional success in replicating itself. The gene, in sum, in its modern, biomolecular configuration, is thought to bear part of the responsibility for its own continued existence, and without this prior recognition of responsibility for its own being, there would be no way to distinguish the gene as such. As Haig puts it:

> Final causes emerge from replicative recursion by the process we call natural selection. Genes and their effects are like eggs and chickens. A gene's *effects* have a *causal* role in determining which genes are copied. A gene (considered as a lineage of material copies) persists if its lineage has been consistently associated with survival and reproduction. If possession of a gene is consistently associated with survival and reproduction then one can infer that the gene's *effects* have *causally* contributed to the gene's persistence. . . . If past effects that have contributed to a gene's persistence are recapitulated as effects of a current gene, then these *effects exist for the sake of their cause* and can be considered the gene's raison d'être (or final cause).[4]

By contrast, an electron cannot be said to be both its own cause and effect, and its legibility, as an electron, does not require recourse to replication, or "replicative recursion."

It may be helpful at this point to return to Walker and Lachmann's distinction between life and aliveness, so that we may see how it pans out for the gene in its informational guise. If a bacterium is both a part of life, being in its lineage, and alive, being metabolically active, and if an electron is neither, then (informational) genes, like textual objects of all kinds, stand between these extremes, partaking of life without being alive. We will address later the reason why metabolism should be considered the sine qua non of aliveness. For the moment, however, it suffices to point out that both bacteria and genes in fact participate in their own replication. They are both replicators (constructors). Yet the different degrees to which they participate in replication has everything to do with whether, and how, each partakes (or does not partake) of aliveness. A poem, although it may not write itself in the same way that a bacterium does, nevertheless, like the gene, does affect the likelihood of its being rewritten within a given ecosystem. The difference between what we consider to be aliveness (in the bacterium) and inanimate forms of life (such as are found in poems or informational genes) consists both in the degree and kind of participation each entity may have in these processes of self-sustenance and self-replication, or what Kant, in short, calls self-organization.

To return to our argument, however we consider the participation of the gene in its replication, and whether or not we consider its role sufficient to qualify it as being alive, the fact that its very legibility, qua gene, requires recourse to a serial process of potentially limitless replication tells us something about the implicit stakes of referring to it as immortal. The gene's immortality, though allegedly an immortality of the dead signifier, being strictly informational, is in fact caught up in and motivated by a prior teleological orientation, one toward the total phenomenon of self-reproduction. When, therefore, we call genes immortal, it is not a neutral claim simply about their potential to survive indefinitely—be it like a stone monument through sheer perdurance (token), or like bits of data encoded on a hard disk through iteration (type)—but, precisely, one that confers agency on these bits, insofar as the data that they encode are ontologically predetermined by their role in self-replication.[5] The immortal gene exists for and by means of itself. The appearance of its neutral informational fecundity is thus belied by teleological notions of autonomy, self-production, and organic wholeness, which invest in it and lend it its specificity. Just as, for Plato, generation was only legible *as* generation by virtue of a prior intimation of immutable self-presence, so too we could say that for the biologist, the gene is only legible *as* a gene by virtue of a preorientation toward reproducibility, which allows this whole to appear as a whole.

The purpose of pointing this out is not to dismiss the significance of either the gene or the unit in which it takes part. It is simply to indicate the artifactual, i.e., non-natural, prosthetic status of the concepts that allow, in each case, for the appearance of its immortality.

In contrast, then, with genetic immortality, which is conceived as essentially informational, metabolic (or biological) immortality is conceived as essentially embodied. It is numerical as well as qualitative; it concerns both the ways in which organisms live and the possibilities of indefinitely extending those lives.[6] To be precise, it concerns how some organisms may live as dying (to adopt Heidegger's terminology), but without ever actually realizing death as an empirical, physiological reality (in demise). In distinction to genetic immortality, then, which concerns less the living as such than it does the text, program, or code that is believed to make life possible, metabolic or biological immortality concerns life as it is lived.

Biological immortality thus concerns life as it is lived. Nevertheless, there is a crucial, if obscure, gap between the minimal, metabolic determination of this potentially immortal way of living and the full, phenomenological sense of world, which is generally associated with all such living beings. It is this gap, which allows for a positive determination of life to coexist with an indefinite determination of organism and world, that constitutes a significant portion of the fascination with so-called biological immortals. If we are to comprehend the uncanny position of the life sciences today, then we must grasp both the nature of this fascination and the nature of the gap through which it is constituted.

The Notion of Biological Immortality

Let us now turn to the mechanisms of so-called biological immortality. How do organisms constitute themselves, and to what extent may death be (for them) a mere physiological contingency? Our first example of a biological immortal is one of the earliest and simplest forms of life: prokaryotic, single-cell bacteria that reproduce through binary fission. In his 1970 book *The Logic of Life,* Nobel Prize–winning molecular biologist François Jacob supplies a typical image. Bacteria, because of their ability to reproduce asexually, need not die. Instead, they may "go on reproducing indefinitely" and only encounter death as a "contingency"[7]:

> The little bacterial cell is so arranged that the whole system can reproduce as often as once every twenty minutes. With bacteria, unlike organisms which reproduce only sexually, birth is not counterbalanced by death. When bacterial cultures grow, the individual bacteria do not die. They disappear as individuals: where there was only one, suddenly they are two. The molecules of the "mother" are distributed equally among her "daughters." For instance, the mother contained a long duplex of deoxyribonucleic acid that splits into two before cell division. Each daughter receives one of these identical duplexes, each of which is formed by an "old" chain and a "new" one.

> One of the criteria that a bacterium is no longer alive is its inability to reproduce. If this non-life is to be seen as death, it is a contingent death. It often depends on the conditions of the culture. When a small part of a culture is continuously replaced by fresh medium, such a culture remains in a state of perpetual growth: bacteria go on reproducing indefinitely.[8]

In order for a bacterium to reproduce, it must first separate the two strands of its DNA helix and replace the nucleic acids corresponding to each. Each of the two newly formed double strands is then allotted to one of its daughters. This process allows the mother cell not only to propagate its genome with fidelity in its offspring but also to create a constant supply of youthful, rejuvenated cytoplasmic material for its lineage. In this way, so long as it continues to reproduce, a bacterium may survive indefinitely.[9]

Single-cell organisms like bacteria thus appear naturally to possess a form of biological immortality: they can reproduce endlessly, without any necessary appearance of death, or self-destruction, for the mother. They preserve themselves for indefinitely long periods of time by continually growing and dividing, and they thereby supply biologists and philosophers alike with a living—and not just hypothetical or speculative—example of biological immortality. As Baudrillard writes: "The evolution of the biosphere is what drives immortal beings to become mortal ones. They move, little by little, from the absolute continuity found in the subdivision of the same—in bacteria—toward the possibility of birth and death."[10]

In his 1949 book *Accursed Share,* Georges Bataille has already explored this alleged continuity between bacterial progenitors and progeny. "As we know," he explains, "death is not necessary. The simplest forms of life are immortal: the birth of an organism reproduced through scissiparity is lost in the mists of time. Indeed, it cannot be said to have had parents. Take for example the doubles *A′* and *A″*, resulting from the splitting in two of *A; A* has not ceased living with the coming into being of *A′*; *A′* is still *A* (and the same is true of *A″*)."[11]

The division of one mother cell into two daughters—a figure of speech that Jacob appears to be comfortable with but that Bataille is not—thus gives us to think an immortal form of life to the extent that death, if and when it happens, happens only contingently. Jacob,

Baudrillard, and Bataille alike lead us to believe that death need not arise within the life cycle of these organisms. Bacteria possess no internal (i.e., genetic or epigenetic) limit on their life-spans but only that imposed by their external environments. As evolutionary biologist and senescence researcher Michael R. Rose puts it, "My view was that biological immortality was found only in organisms that reproduce by splitting symmetrically in two, like many bacteria or sea anemones. . . . Most fissile animals are small and vulnerable. But they do not age. For this reason, they are conventionally called 'biologically immortal.' "[12]

Biological immortals are those organisms that may grow, develop, and reproduce over an indefinite span of time and/or generations, without succumbing to death. Generally speaking, this condition coincides with a resistance to aging, but strictly speaking, it need not. The prototypical case of a biological immortal is the fissile bacterium, but many clonal unicellular and multicellular eukaryotic organisms also attain to biological immortality, although not in the same fashion. As I shall explore below, crucial differences emerge between the forms of biological immortality available to morphologically diverse life-forms.

Thus, in distinction to the ideal of immutability that marks the thought of the immortality of the gene, there is without doubt change within the life cycles of asexual unicellulars. Like all life, the bacterium does not perdure immortally according to Diotima's or Descartes's, or even Dawkins's, Dennett's, or Haig's, ideals of stasis and stability. As Jacob points out, the price paid for this eternal life is the disappearance (but evidently not death) of the individual, for where there was one, now there are two (or, in certain cases, even more). Nor are each of the daughters produced in this way simply identical, either to the mother or to one another. Studies have documented that as part of their natural cycle of growth and division, bacteria can introduce phenotypic variation among offspring as a response to various environmental factors.[13] Moreover, with the production of each successive generation, genetic variation inevitably arises. Each time a bacterial cell forms a new set of chromosomes and then divides, copying errors can occur. A truly infallible process of genetic replication, whose fidelity would be 100 percent, is neither conceivable nor desirable from the Darwinian evolutionary perspective shared by the majority of biologists today. To speak of the biological immortality, or indefinite life, of the bacterium is thus to speak of a form of survival that ought to persist in spite of the following:

1. Inevitable alterations to its genetic code which, sooner or later, may lead to morphological changes and even events of speciation across generations.
2. The constant metabolic transfer of matter and energy between each bacterium and its environment.
3. Ongoing processes of growth, repair, and division that, under optimal conditions, leave each bacterium recognizable in organization but not in the material components that make up its particular structures (i.e., there may be qualitative, but not numerical, identity of individual elements of its anatomy).[14]
4. The perpetual possibility of "sexual" contact with other bacteria, in horizontal gene transfer.
5. The constant disappearance of the individual in the formation of two (or more) autonomous progeny.[15]

For all these reasons, biological immortality—or potential immortality, as it is sometimes known—is not an uncontroversial assignation. Indeed, it was already a subject of controversy in the late nineteenth century. "The process of fission in the Amoeba has been recently much discussed," writes Weismann in an 1881 lecture entitled "The Duration of Life,"

> and I am well aware that the life of the individual is generally believed to come to an end with the division which gives rise to two new individuals, as if death and reproduction were the same thing. But this process cannot be truly called death. Where is the dead body? What is it that dies? Nothing dies; the body of the animal only divides into two similar parts, possessing the same constitution. Each of these parts is exactly like its parent, lives in the same manner, and finally also divides into two halves. As far as these organisms are concerned, death can only be spoken of in the most figurative sense. There are no grounds for the assumption that the two halves of an Amoeba

> are differently constituted internally, so that after a time one of them will die while the other continues to live.[16]

Though Weismann, much like Jacob after him, ultimately concludes that fissile organisms are immortal, it is only after admitting that the whole process defies our common understanding of organismal individuality. In both cases, it is more the absence of a "process [that can] be truly called death" that leads each biologist to this determination than it is the presence of any robust notion of individual bacterial, or amoeboid, survival. Biological immortality—so far as it applies to single-cell organisms anyway—thus appears to imply a continuity of the processes that constitute life, without, however, indicating anything about the subject of those processes. Yet even this alleged continuity, or absence of death, has recently been brought into question:

> The determination of mortality and immortality is even more complex in a stricter biological sense. A common example given for biological immortality is the alleged indefinite replicative capacity of unicellular organisms because cadavers do not result from fission replication and an endless lineage of viable cells are produced. However, this example fails to distinguish between individual and population mortality and immortality. Individual micro-organisms, like individual humans are mortal, but as populations both appear to be immortal. The "cadavers" of organisms that divide by fission are their molecules and atoms that become increasingly diluted and ultimately lost in those cells at the leading edge of the expanding population. Or, these molecules are lost to further progeny when fission products die.[17]

The commonly accepted immortality of unicellulars is thus not without its difficulties, and the root source of these difficulties, as we shall see, is both empirical and philosophical in nature. According to microbiologist Leonard Hayflick, individual microorganisms are no more immortal than are individual humans because in both cases, individuals are subject to death.[18] It is only "as populations" that the former—or the latter—can be described in this way. The mistakes of those such as Weismann, Jacob, and Rose would thus result, according to Hayflick, from a misinterpretation of the signs that constitute individual, microbial death.

Hayflick's rebuttal raises a number of questions. Leaving aside for the moment the legitimacy of his comparison between populations of microorganisms and populations of macroorganisms—to which we will come—Hayflick's contentious claim concerning the presence of death among unicellular clades seems to be in direct conflict with all previous assertions of their immortality. Yet even for Hayflick some degree of immortality would nevertheless persist for these organisms, but only insofar as they exist as populations. The question thus becomes, in what consists the difference between an individual and a population of fissile organisms? As I will argue, the pertinence of this distinction for unicellulars is not as obvious as it initially appears, especially when it is used in reference to their so-called biological immortality. Hayflick believes that he can speak of the death of individual microorganisms and the life of populations. Moreover, in doing so, he presumes that the life or death of the individual actually matters for the kind of immortality that is generally ascribed to them, be it by Weismann, Jacob, or others. But what if it doesn't matter? What if the presumed absence of death among bacterial cultures has always been a red herring?

Biological Immortality versus Personal Identity

The apparent conflict between Hayflick and these others comes down to the nature of the identity that is supposed to be preserved in division. This identity must be strong enough to constitute sufficient grounds for the claim to immortality. If it is not, then we should simply abandon the term and speak instead of the more familiar phenomena of reproduction and birth, each of which implies a substantive difference between progenitor and progeny. There are, however, multiple kinds of identity; the difficulties that arise whenever one speaks of the reproduction of unicellular organisms result from the conflation of an identity, first, of sheer biological or metabolic continuity, through which any organized, biological structure can maintain itself as an organized, biological structure, with that, second, of a personal or individual nature, through which any being can project itself as a being, in its world. Though these two determinations of identity are in practice always interconnected, they are not simply coconstitutive. One does not simply entail the other. The difficulty—but also fascination—with immortal unicellulars results from the inherent ambiguity of this interconnection.

It is only from the latter perspective—that of personal or individual identity—that one can ask whether the mother bacterium is preserved in all its offspring, or just one. While questions such as this one are probably inevitable, they are also self-defeating. If the mother is preserved in only one daughter—for example, the lineage that in each instance receives the grandmother's "original" strand of DNA—then it is both unlikely that anything like immortality is actually possible—as Hayflick points out—and unclear why this should be privileged as the true seat of bacterial, or amoeboid, ipseity. Thus, identifying a single inheritor after each event of fission, which appears to be necessary if we are to preserve a meaningful notion of individual identity, is an at best arbitrary gesture in the case of bacteria. If, however, the claim is instead that the mother is preserved in all its progeny (i.e., in the population), then we avoid Hayflick's initial criticism, but in so doing, we are forced to explain by what rights one can become two, then four, eight, sixteen, thirty-two, and so on, all while remaining one. In brief, the concept of individuality would somehow have to be squared with duplication.

Branching problems such as this one have been well rehearsed by Derek Parfit and others. In short, just as soon as one adopts the position of the individual, one either loses sight of the problem altogether, or one must start speaking of disappearance without death, which clearly does not clarify anything. When one treats the problem of unicellular division from the perspective of personal identity—that is, beginning from a mother—then paradoxes that arise from all such branch-line cases are inevitable.

If we are to explain what happens in division and clarify the quandary of the biologists, then the first crucial step lies in establishing that baseline biological immortality refers only to a physiological condition. It denotes metabolic continuity, but it does so without implying anything about what we commonly think of as individual, or personal identity. The immortality of branching bacteria, amoeba, protists, and archaea is therefore founded on different grounds than the immortality that we commonly refer to when we speak of any being concerned with itself—concerned with an "itself"—for which branching poses a devastating impediment. Whether or not, then, bacteria or amoebas possess a sense of self, and whatever happens to that sense, in division, the biological concept of immortality by definition must remain agnostic. All that this concept is concerned with is whether, physiologically speaking, there is continuity of process—that is, metabolism. As Parfit puts it, in speaking of division, "Whatever happened to me, we could not discover what happened."[19]

From the biological perspective that is Weismann's, Jacob's, and Rose's, the key to the survival of unicellulars consists in the ongoing presence of metabolic processes, the persistence of the genome across events of division, and the evident absence of death for the individual.[20] The problem with their line of approach is that just as soon as one adopts the position of biological immortality, one loses any right to speak of individuals. But that is not all, for as a consequence of this loss, the very difference between the individual and the population also begins to break down, and it is for this reason that we cannot simply subscribe to Hayflick's position either. The difficulty, as we shall see, consists in determining the meaning of biological immortality in the wake of its voidance of the individual (subject). Biological immortality appears to be meaningful, but it cannot be so in any way that could possibly correspond to the forms of continuity, identity, and individuality familiar (and important) to us. Biological immortality is, in the final analysis, strange. But being strange, it cannot but be imbued with concepts that ought to be derivative with respect to it. Our fascination with it testifies to this projection and displacement.

Thus, insofar as the term "biological immortality" can be applied to single-cell organisms, it can in no way imply anything about any internally produced bacterial or amoeboid ipseity. It says nothing about the individual. It refers only to a continuity of physiological functioning, paired with a high degree of preservation of the genome. (Without this second factor, there would be no way to differentiate clonal from sexual reproduction, and thus there would be no way to oppose the continuity of unicellulars from the reproduction of sexual multicellulars, such as *Homo sapiens,* which always involves preservation of metabolism at a single-cell level.) Metabolic and genomic continuity is thus what biological immortality should mean. To this extent, it is an entirely objective determination, one based on principles produced by the life sciences alone—which is to say that it has its truth, but a truth only with respect to the limits set by these disciplines.

Where trouble starts to emerge is when this form of physiological continuity is inevitably made to face, and is then invariably conflated with, other forms of continuity common to organic life-forms, such as individual, organismal, or personal identities.[21] As should by now be clear, biological immortality implies the most radical departure from these conceptions. Yet it is precisely the position of personal identity that Hayflick adopts when he presumes that there is a difference between an individual and a population of microorganisms. Such a difference surely applies to *Homo sapiens* just as soon as individuals begin to declare their identities as individuals. It might well apply to bacteria as well, were we better able to read their signs of life, and were we concerned with these signs, when speaking of their immortality. The issue is that insofar as the concept of biological immortality is concerned, both cases are irrelevant. Biological immortality does not speak to any instance of personal, internal, organismal, or phenomenological identity. All it testifies to is the (indefinite) persistence of metabolism, be it within or across bodies. Indeed, its very conceptuality testifies to the indifference of this difference. In this respect, the demise of one member of a clade is immaterial to the survival of the whole, each member of which maintains physical and physiological continuity with the mother, and the mother's mother, and the mother's mother's mother, even if there is also division, transformation, conjugation, and even speciation across time. But so far as biological immortality is concerned, there is no individual, no ground on which to determine the difference between members of a clade. Indeed, to speak of members of a clade is already to miss the essentially branching structure of this form of vitality and to appeal to another measure, or heuristic, altogether.

Thus, even if, contrary to Hayflick, there were no death among unicellulars—that is, no loss of molecules and atoms—it would make no difference to the kind of immortality that is ascribed to them, for there would still be no individual immortality because biological immortality is blind to the individual. Its conceptuality requires that we read events of division as nonevents. There is only a single, continuous, and indefinitely expanding stream of the living. In this respect, the case of unicellulars is not one among others; rather, it is the source from which all future determinations of physiological continuity take shape. The example of minimal autopoietic survival and division is not just an example. It is the founding instance of all biological continuity, and one that persists across all known forms of life, preserved in both the simplest and most complex organisms, thereby allowing scientists to unify the totality of the biosphere under a single, ever-present, and living umbrella. This is why it is essential that we grasp both what it entails from a technical standpoint and what it means from a philosophical one.

The immortality of unicellulars is thus the basic fundament of the vitality of life itself. This is perhaps why there is so much pressure to see it—that is, to read it—as a form of continuity and persistence, even if, as is all too obvious, the transformations that single-cell organisms necessarily undergo in division are already a form of death. Division—what Jacob calls the disappearance of the individual—already entails the necessary possibility of the end of ipseity (death) in the branching of one into two (or more), just as horizontal gene transfer does. This too is a form of death. What it is not, however, is death in the physiological sense of the word. The death of self cannot, however, simply be equated with that other form of death—of empirically identifiable physiological life. Of this, the case of unicellulars is the exemplary instance. Nevertheless, if Hayflick is guilty of conflating what are in fact two distinct senses of immortality and death, it is neither simply by accident nor through mere neglect. Insofar as metabolism implies cognition, as we shall explore in chapter 13, biological immortality will always have some relation to ipseity. It will always be caught up in problems of identity and individuality. The very recourse to the term "immortality" is perhaps already sufficient to testify to this complicity, for if there were not already some intuition of a preserved self, abstracted from the evident absence of physiological death, then why would one even bother appealing to it? Does not the enduring presence of the term "immortality," even and especially in light of the biologist's uneasiness toward it, speak to this persistent allure?

Now we must ask what is at stake in the determination of this form of biological continuity (immortality). What is invested in this notion that, we have just seen, can only be maintained to the extent that it is divorced from every concept of self?

As developmental and synthetic biologist Michael Levin explains, in speaking of planarians, which enjoy a life cycle of clonal propagation similar to that of unicellular organisms:

> [Flatworms are] immortal, they don't age, there's no such thing as an old planarian, which means first of all that the worms that we have in our lab are in a direct physical continuity with worms that were here half a billion years ago, and that's mind-boggling to think about, and also it's telling us that in fact, it is possible to be a complex organism with a good brain and so on and never die.[22]

Similarly, neurobiologist Antonio Damasio notes:

> In a simple way, the region of unlikeness called life, at the level of humble cells—without and with a nucleus—or of large multicellular organisms such as we humans are, can be defined by these two traits: the ability to regulate its life by maintaining internal structures and operations for as long as possible, and the possibility of reproducing itself and taking a stab at perpetuity. It is as if, in an extraordinary way, each of us, each cell in us, and every other cell were part of one single, gigantic, supertentacular organism, the one and only organism that began 3.8 billion years ago and still keeps going.[23]

In an equally radical version of the same thought, astrobiologist Sara Walker provocatively puts it thus:

> The first cell never died, it bifurcated in time, generated all biology, all of us, and all technology. Life is a lineage of events—everything we do becomes a part of that lineage.[24]

Finally, in the words of Lynn Margulis:

> In modern language, we can say that all organisms alive today or in the past share a physical continuity with all others. All organisms are part of a single continuous bounded autopoietic system that has never been breached since the origins of life in the Hadean or Archean eon.[25]

In what lies our fascination with a being, at once singular and plural, seemingly able to extend itself without limit, and even, to all appearances, to do so without difference, across all fathomable stretches of space and time? Whence the fascination with such an example? If the form of immortality particular to unicellulars ultimately implies nothing about individual identity, as I have been arguing, then it nevertheless remains the case that in our desire to locate in them a form of sheer physical and temporal continuity, something more than mere objective knowledge is at stake. From the example of the individual prokaryotic cell, one quickly escalates not only to all eukaryotic and multicellular organisms but also to the totality of life (and technology) itself. The issue is not that such extrapolations are hyperbolic. They are not—or, anyway, what is hyperbolic is not the extrapolation but the principle that undergirds the initial, exemplary instance. Just as soon as one accepts the continuity of the single-cell prokaryote that divides through fission, the rest follows, almost as a matter of course. Yet what the extension of the continuity originally found in the minimal instance of the single cell, up through all of life and all of technology reveals, are the stakes at work in that primary example.

Our investment in the phantasm of what Baudrillard calls the "pure contiguity" of dividing single cells goes to the heart of our investment in life itself.[26] Biological immortality may not imply the presence of organismal identity, but just as crucially, neither does it preclude it. Instead, it announces a site of deep, evidently empirically identifiable continuity, right on the organism itself, and thus right on life. What this continuity and contiguity permit, in turn, is the very possibility for lineage—now in the strong sense of the term—as the indefinitely extended, unbroken line. The existence of life today evidently demands belief in a continuous, 3.8-billion-year-old succession.

Yet the question that all this raises is whether our investment in continuity and contiguity precedes or follows investigation into the phenomenon of life. Is not our very fascination with the phenomenon in some sense preoriented by this investment? Why else would we seek to explain the lineage of life itself *as* a lineage? In other words, why else consider all of known life as but a single example—as the astrobiologists are so fond of saying—rather than as the barely comprehensible, intimately connected, but ultimately differentially, and thus discontinuously, related concatenation that it also is?[27] What is at stake in the determination of the immortality of unicellulars? Well, as I have been arguing, nothing short of the nature of life itself.

Already in chapter 4 we encountered an example of the undifferentiated continuity that cellular fission evidently allows us to think. For Nancy, the form of life embodied by cancerous growth was just such an instance, even if it was one that had to remain separate from the difference constitutive of "existence" proper: "Without death there would only be contact, continuity, and contagion, a cancerous propagation of life that would consequently no longer be life—or rather, it would only be life and not *existence,* a life that would not at the same time be *anastasis.*"[28] What is crucial in passages such as this one is that even as he spurns this figment of pure life, Nancy's opposition of "existence" to "cancerous propagation" is nevertheless sufficient to uphold the latter as a possibility—a possibility that may alternatively be embraced or scorned. The question, raised no less by Nancy than by Margulis, Walker, Damasio, and Levin, is this: what, after all, is sheer contact, contiguity, or continuity? Where do these concepts, ideas, or notions come from in the first place? Do the examples of cancerous growth, bacteria, and amoebas really produce novel concepts of contact whereby they are interpreted, or does our empirical, scientific understanding remain reliant in each case on derivative transpositions of metaphysical notions of contiguity? However one responds to these questions, the following point remains: the fact of our fascination with such examples of biological immortals testifies that they may hold some promise of sheer contiguity for us.

If, as I have argued, the ascription of biological immortality must ultimately remain agnostic with respect to selfhood and individuality, being a strict description of continuity of process, then whence comes the investment in this identification, and this continuity? Biological immortality may not entail organismal identity, but neither does it preclude it. Instead, it holds out a space of sheer potential, as though the latter might then come to occupy it. But the lie consists in the deception that this pure contiguity, direct physical continuity, and undead, ever-present first cell—of which we, all of us, and all of our technology, remain a part—would be conceivable without our first having access to the kernel of sameness that ipseity inaugurates and names. In other words, the lie consists in believing that a strictly empirical determination of contiguity and continuity would be possible apart from, and without already being informed by, our experience of these forms of presence, concerning which readings in Plato and Derrida should by now make us quite wary.

The purpose of raising the question of the immortality of unicellulars has been less so that we could prove or disprove its existence, on the standards of the biologist, than it was to reveal the central conceits that inform basic notions of biological continuity. Unicellulars that reproduce through fission present difficulties for anthropocentric (or, more broadly, metazoan) notions of life, death, and self. However, even if branching microorganisms ultimately short-circuit our best conceptual apparatuses for conceiving of them, and even if we know, deep down, that these organisms only persist by virtue of the difference, in division, that constitutes them, they nevertheless remain the first and last instance of stability amid the endless change and mutability that marks the physical world. Thus, in a perverse twist, it is to the extent that biological immortality remains agnostic toward individual identity that it can become the vehicle for sustaining its possibility. Even if the empirical determination of metabolic continuity can never fully embrace this possibility of selfhood, neither can it ever wholly preclude it. By being invested with the telos of self-presence, it maintains its promise by upholding a categorially indefinite relation to the being of the beings it reads.

10

INTERLUDE ON DEATH, DIVISION, AND THE LIFE OF HENRIETTA LACKS

> If allowed to grow uninhibited under optimal cultural conditions, [HeLa cells] would have taken over the world by this time. . . . The biopsy . . . has secured for the patient, Henrietta Lacks as HeLa, an immortality which has now reached 20 years. Will she live forever if nurtured by the hands of future workers? Even now Henrietta Lacks, first as Henrietta and then as HeLa, has a combined age of 51 years.
>
> —Rebecca Skloot, *The Immortal Life of Henrietta Lacks*

> That lady has achieved true immortality, both in the test-tube and in the hearts and minds of scientists the world over, since the value of HeLa cells in research, diagnosis, etc., is inestimable.
>
> —J. Douglas, "Who Was HeLa?"

The problem posed by the immortality of the bacterium—a unicellular organism that changes and multiplies but evidently does not die—forced us to confront what in our notions of identity and difference seemed inadequate to its mode of being. The bacterium may indeed be immortal, but the price paid for its longevity is a multiplication of bodies that hardly seems adequate to the ideal of individuality that underlies the designation in the first place. That is, anyway, if the designation is to remain meaningful.

If the bacterium truly poses a problem for these concepts, then it is not because it leaves them intact while simply resisting their grasp; rather, it is because it implicates them in its own vital life. The problem with the bacterium is that it calls to our notion of immortality while remaining resistant to it. In the choice between simple ascription and simple dismissal, both options seem inadequate.

The purpose of these observations is not to draw any direct analogies between the lives of unicellular prokaryotes and those of multicellular, eukaryotic, or minded mammals. It is instead to begin to unravel a problem of formal organization that will be reproduced at different scales across the biological world. The temptation to identify bacterial immortality is not simply a function of anthropomorphism; nor is it only a misguided artifact of poorly formed biological concepts. It is bound up with the most basic conceptions of life and inheritance, where they remain entangled with the intersection of biology and culture; and it is, to this extent, neither to be embraced nor dismissed.

In the present chapter, I want to explore an example of this temptation, and both its promise and peril, with the story of Henrietta Lacks. Lacks was an African American woman who lived in Baltimore, Maryland, and developed cervical cancer in the 1950s. During her treatment at Johns Hopkins, samples of her cancerous cells were taken from her cervix and surreptitiously kept alive in the lab of Dr. George Gey. Remarkable for their capacity to survive and reproduce, seemingly without limit, Lacks's cancer cells, known today as HeLa cells, would become "the world's first immortal human cells."[1] That is, they would become the first human cells to be grown indefinitely in vitro. Their singularity, which has contributed to their being cultivated in laboratories across the world, consists not only in the speed at which they grow and their ability to survive harsh conditions, but also, just as remarkably, in the total lack of senescence that they demonstrate.[2]

Lacks's cancer cells have thus been judged to be immortal in precisely the sense that bacteria, amoebas, and other unicellular fissile organisms have traditionally been: they grow and divide, seemingly without end, all while maintaining their physiological and anatomical organization across generations. As Rebecca Skloot, author of *The Immortal Life of Henrietta Lacks,* explains:

> Gey and his wife, Margaret, had spent the last three decades working to grow malignant cells outside the body, hoping to use them to find cancer's cause and cure. But most cells died quickly, and the few that survived hardly grew at all. The Geys

> were determined to grow the first immortal human cells: a continuously dividing line of cells all descended from one original sample, cells that would constantly replenish themselves and never die. Eight years earlier—in 1943—a group of researchers at the National Institutes of Health had proven such a thing was possible using mouse cells. The Geys wanted to grow the human equivalent.[3]
>
> [The cancerous cells grew in culture] twenty times faster than Henrietta's normal cells, which died only a few days after Mary put them in culture. As long as they had food and warmth, Henrietta's cancer cells seemed unstoppable.[4]

As a result of the work of the Geys, HeLa cells thus became the world's first standardized human cells available for medical and scientific experimentation. Yet they have also become the subject of great controversy—not only as a result of the way they were taken, but also because of the relation they maintain to their source. The biological immortality that makes them of so much scientific interest also guarantees that they preserve a form of living, metabolic, and genetic continuity with Lacks, their originator. Both their medicoscientific interest and their ancestral pull, qua the lineage of their descent, are, in other words, possibilities opened by the forms of continuity undergirding their biological immortality. Thus, as I will now explore, the life of Henrietta Lacks—which is to say, the immortal life of Henrietta Lacks—is noteworthy to the extent that it demonstrates the bind surrounding indefinite metabolic survival, which, being tied to basic biological notions of life and presence, cannot but hold out the promise of something more.

The bind consists in the following: HeLa cells both are and are not Henrietta Lacks. They both partake and do not partake in the life of Henrietta Lacks. Henrietta Lacks both survives and does not survive in the millions upon millions of cells that live on in laboratories across the globe. To say this is not simply to embrace paradox at the expense of common sense. Nor is it merely to conflate two distinct types of survival, which might otherwise be parsed. It is rather to acknowledge the mutual source, on the one hand, of the conflicted experience of her legacy in the eyes of those who bear it, and, on the other hand, of the biological notion of continuity that operates in, even if it does not exhaust, the former. It is this essential complication of biology and identity that marks our biotechnological modernity. Biological science cannot adjudicate the limit between organic nature and personhood. It helps to establish this relation, but in doing so, it cannot shore up its border.

The entangled legacy of Lacks with HeLa is, then, if not *the*, then *a* central motif of Skloot's book, and a difficulty with which it constantly grapples. In order to demonstrate this legacy, I cite variously from her text:

> Deborah believed Henrietta's spirit lived on in her cells, controlling the life of anyone who crossed its path. Including me.
>
> "You're writing a book about my mama. . . . Her cells grown big as the world, cover round the whole earth," he said, his eyes tearing as he waved his arms in the air, making a planet around him. "That's kinda weird. . . . They just steady growin and growin, steady fightin off whatever they fightin off."
>
> Henrietta's cells were still alive.
>
> How come nobody told her family part of her was still alive?
>
> Part of your mother, it's alive!
>
> They said they got my wife and she part alive.
>
> But Deborah couldn't stop worrying. She was terrified that she might have cancer, and consumed with the idea that researchers had done—and were perhaps still doing—horrible things to her mother.[5]

HeLa cells both are and are not Henrietta Lacks. This is what one hears throughout the testimony of her surviving family. But what exactly does this mean? The purpose of citing Lacks's kin in this way is not to affirm any singular interpretation of the relation between "Lacks" and "HeLa" but rather to witness the power of the relation that persists through these cells as an enduring question. Nearly everyone Skloots interviewed continues to be haunted by the possibility that Lacks remains, in some way, in HeLa cells. If at times this manifests as a distressing form of personification—as if Lacks herself still felt through her remaining cancerous cells—then the question becomes why such an identification is even possible. The relation between "Henrietta Lacks" and "HeLa" persists as an open question. It persists in this way on the one hand because of the principle of metabolic continuity/contiguity, just unearthed in chapter 9, and the essentially indefinite relation between this continuity and personal identity. On the other hand, it also persists because metabolic continuity, for all its problems, in fact performs a form of contact that, even if clearly of a different nature from personal identity, is nevertheless from a bioempirical perspective undeniable. Something, therefore, is preserved. Whether one identifies this something with "Lacks" or not, therefore, does not come down to anything empirically determinable; instead, it comes down to where one first situates the essence, or limits, of the person. The ultimate irony and hypocrisy of biomolecular understanding would be to try to dismiss, on medicoscientific grounds, the meaningfulness of the metabolic continuity that it had attempted to establish as a primary biological relation.

The testimony of Lacks's family thus stands as the most pointed example of the irresolution of this bond, as it (evidently) passes from the medicoscientific to the interpersonal realm in what would be a transition from the empirical to the cultural, if not for every evidence that the empirical notion of continuity is already seeded by cultural investments in, and conceptions of, contact and presence. *The Immortal Life of Henrietta Lacks* contends with the legacy of the open question that is the relation between biological function, subjective identity, and material inheritance. That is the point. Lacks's survival remains an open question, and one not susceptible to simple resolution. A few comments about this are necessary, as follows.

1. To say that Henrietta Lacks may survive in HeLa cells is neither to embrace genetic reductionism nor to discount other viable modes of determining identity and survival, including those that would discount this one. Rather, it is to acknowledge that one could not simply dismiss the partial identifi-

cation of Henrietta Lacks with HeLa cells on some kind of objective or universal ground. There will always remain the possibility of identification based in the physiological, anatomical, and genetic continuity that persists between these cells and those of Henrietta Lacks, from which they trace their ancestry.

2. To say that Henrietta Lacks may survive in immortal HeLa cells is not to say that anyone, *de son vivant,* would either prefer or even accept this form of immortality as sufficient to what one says that one wants when one says that they want to live forever. That desire for self-perpetuation, to which we will yet return, need not be satisfied by the present iteration in order for us to acknowledge that HeLa cells nevertheless must be accepted as a living trace of Henrietta Lacks. To deny this would be to deny not only the experience of those who bear her legacy, which experience offers, at minimum, a testimonial force, but also the very biophysical basis of individuality, and with it one of the few empirical holds we have on personhood: the spatiotemporal continuity of bodies. Again, the purpose of pointing this out is neither to somehow equate HeLa cells with Henrietta Lacks nor to reduce personhood to mere biological autonomy. It is rather to acknowledge that even if a veritable abyss separates the Henrietta Lacks who lived, willed, suffered, and died from the cancerous HeLa cells that survive her to this day by growing and dividing without end, nevertheless, Henrietta Lacks survives in them. If it is a function of the specificity of our genetic modernity that such a link becomes legible in the first place, such that we could not universalize it and could even imagine a set of biological and juridical principles that would dismiss it, then it is no less the case that any identity that replaces this one must itself find its ground in turn. It is perhaps worth noting that when Gey himself was on his deathbed, terminally ill with pancreatic cancer, he too wished for a sample of his cancerous cells to be taken and cultured, so that he too might somehow live on, if only partially, in them.[6]
3. Finally, the analogy between cancerous growth and bacterial reproduction can only be partial. Whereas bacterial

> immortality is of the whole bacterium, in the case of HeLa cells, we have only a partial survival of the original, differentiated being. We will return to this problem in the chapters that follow.

In reading *The Immortal Life of Henrietta Lacks* and encountering the pain left by the uncertain presence of Lacks in HeLa cells, what we thus find is an expression of the problematic logic of identity itself. What the case of Henrietta Lacks announces, in its fashion, is the unresolved bond that subsists through differential relations of identity, which are neither simply present nor simply absent. Thus, I do not believe that one should simply affirm such relations, or simply deny them. One would be no more justified in claiming that Henrietta Lacks lives in HeLa cells than one would be in denying that she does. Different grounds can support either claim. What strikes us in statements such as "Henrietta Lacks is immortal" is therefore less the failure of the claim, on the basis of a more real or legitimate standard of immortality, than the necessity of admitting its possibility, and on the basis of this possibility, of further admitting that all immortality has always been a question of translation, even and especially that which affirms the pure presence of the original and the undifferentiated nature of what lives on.

Biological immortality, in the sense of sheer indefinite physiological continuity, is not, in and of itself, sufficient to speak to the survival of a self. Be that as it may, what Lacks's case shows is that even if it is not sufficient for selfhood, the perdurance of physiology, by way of a continuous historical lineage, doubtless speaks to something. It speaks to the persistence of a part, whose participation in a whole cannot simply be discounted or explained away.

11

BIOLOGICAL IMMORTALITY II

Multicellular (Metabolic)

The dividing zygote is a clone. Just like a clone of unicellular organisms, embryonic cell divisions pass on any generic variation which has arisen in previous divisions.

—Leo Buss, *The Evolution of Individuality*

Asexual reproduction propagates clones, which often remain attached to the parent plant and can proliferate to establish community-sized "individuals" of extraordinary longevity, maybe in excess of 10,000 years. In this respect, clonal plants resemble the huge underground hyphal networks of certain fungi, some of which may be even older. . . . Clonal behavior like this really does stretch the concept of organismal individuality beyond breaking-point.

—Howard Thomas, "Do Green Plants Age, and If So, How?"

It is possible to argue that there are plant species that are effectively immortal. It is thought that some grass clones have survived more than 15,000 years. . . . In such species, a given clonal genotype, called a *genet,* may be represented by numerous interconnected plants, each called a *ramet.* The ramets may die and be replaced, but the genet survives indefinitely. This form of life is like that of coralline coelenterates, in which coral aging is very difficult to establish, even though the zooids may exhibit obvious signs of aging.

—Michael R. Rose, *Evolutionary Biology of Aging*

Scaling Up

The most ubiquitous form of biological immortality is to be found in single-cell organisms that reproduce through fission. These organisms are capable of dividing indefinitely, which they do without either great mutation or senescence.[1] A certain continuity between generations—that is, the persistence of both organic matter and function—paired with an inability to identify a clear event of death—or, more precisely, a full cessation of the metabolic processes of the parent organism—thus leads to the determination of the immortality, or indefinite life, of these life-forms. Such an immortality does not consist in immutability. Nor does it imply anything about the being of individual organisms within a given lineage. It consists rather in a life cycle without clear, internally driven (i.e., genetically or epigenetically determined), empirically identifiable event of death.[2] A relatively stable, even if duplicated or branched, metabolic being is thus perceptible over time.[3]

As we have already observed, difficulties begin to arise when this form of minimal, metabolic survival is inevitably set against, made to speak to, or held accountable for so-called higher order notions of identity. This happens just as soon as one attempts to establish the meaning of the immortality of single-cell organisms. As Levin, Damasio, Walker, Margulis, and Hayflick suggest, it seems that the example of fissile unicellulars invariably leads to one of two interpretations: either all of life as we know it, descending from a last universal common ancestor, is part and parcel of the same, ever-morphing, immortal cell, or none is. No empirical evidence can decide between these alternatives, because once life is defined metabolically, it cannot but lead to the former conclusion, and once defined organismally, to the latter.[4]

The biological immortality (or indefinite life) of single-cell organisms is thus made possible by a specific set of conditions that must be distinguished from those that inhere in other orders of organisms—notably, in more complex, differentiated, multicellular life-forms. This is true even though all known forms of life in fact preserve, at the level of the single cell, survival by simple fission (or what for our purposes here amounts to the same thing: survival by mitosis). This chapter and the next—chapters 11 and 12—will explore some of the other forms of biological immortality that emerge for multicellular organisms as we build up to an analysis of current projects of radical life extension.

When we go up a grade on the level of complexity, there are at least two known biological pathways available for multicellular organisms to partake in forms of perdurance at least somewhat comparable to what bacteria, archaea, and protozoa enjoy. These two pathways include vegetative propagation—that is, cloning, or asexual reproduction—and indefinite aging, either through nonsenescence or desenescence. While cloning largely reproduces the paradigm of metabolic immortality already identified in single-cell organisms (with the primary difference being what divides no longer does so cleanly because the organism is now constituted by a differentiated, multicellular whole, while the preserved portion can be as minimal as a single cell), indefinite aging inaugurates a new model with novel implications. With indefinite aging, we encounter, for the first time, the possibility not simply of metabolic but also of individual immortality, which is the first step in the direction of something like personal immortality. Individual immortality through indefinite aging thus preserves not only biological function but also, possibly, the being of the organism as an organism. At any rate, this is its conceit.

With indefinite aging, we thus begin a new stage of our discussion of forms of biological immortality, and one with even greater significance for projects of radical life extension. Indefinite aging seems to be the most intuitive and least controversial way of attaining indefinite life for *Homo sapiens* (that is, if such a thing were possible, given human physiology.) It appears to be the most straightforward solution to the threat of death, a mere continuation of embodied individuality as we already know it—save, of course, for the genetic tweaks and pharmacological supplements that it will take to extend the upper limit on our health spans. Although it theoretically appears to be the most straightforward solution, in practice, it is not always so easy to tell where mere metabolic survival ends and full-fledged individual vitality begins. As we shall see, this difficulty arises from the very nature of organismality. Once we start attending to the fluid limits of individual organisms, such individuals can become hard to square with the sort of stability generally expected from them, above all if they are to be considered immortals.

To be clear, this general instability would not be an issue—it is not as though we encounter complaints from the organisms themselves!—if it were not the case that contemporary projects of life extension rely on a link between biological organization, individuality, personhood, and subjectivity. Far from rejecting such a link, however, what I would like to question is whether the alleged biological basis of these identifications is any more or less secure, any more or less performative, any more or less a matter of belief or faith than the prescientific prostheses of immortalization such as poetry, memory, and procreation that these contemporary life-extension efforts are tasked with replacing.

I thus begin with the possibilities of survival opened by cloning before turning to those inaugurated by nonsenescence and desenescence. Along the way, we will have to contend with the nature of organismality. At stake in the presence, or absence, of the organism is also the difference between mere metabolic and individual immortality. If, in the final analysis, cloning still poses some enduring temptation for us as a culture, then it perhaps results from the insecurity of this border.

Clonal Propagation, Indefinite Aging, and Contextual Organismality

If single-cell organisms must divide in order to persist, whereby they are always either multiplying or dying, then, in similar fashion, many plants and fungi reproduce through a form of self-cloning, whereby a portion of the original, such as a spray or graft, breaks off and grows into a new plant. This new plant is not only genetically identical to the parent but also, like daughter and mother bacteria, maintains a degree of physical (i.e., anatomical and physiological) continuity with it. This form of asexual or vegetative propagation involves no element of genetic shuffling, which is why it has been used by cultivators for thousands of years. The only variation introduced into the generations of plants produced through cuttings is what might emerge internally in each one, as its individual cells divide for the purpose of growth and slowly introduce copying errors that, if unaddressed, may become mutations.[5]

Like bacterial reproduction, then, vegetative propagation involves the persistence of a genetic identity between parent and child organisms (allowing, of course, that there can always be mutations), as well as physical and functional continuity (the fractured portion of the parent plant that regenerates into a new plant must remain metabolically viable at the minimal level of the single cell). Unlike bacterial reproduction, however, vegetative propagation does not use up all of the parent. There is visible loss and thus often an undeniable presence of death. An individual apple tree can live, on average, anywhere between thirty-five and a hundred years, after which point it perishes, even if, during its life, it produced many scions. With few exceptions (to which we shall come), the individual plant specimen that clones itself does not survive indefinitely. Only the series of clonal relations—or clone mates—that it produces does. Mortality of the individual plant is thus at best counterbalanced by two things: the indefinite life of the germ line, which is propagated in each of the parent plant's clones; and the life of a "new" plant or many new plants that not only share in the parent's genetic identity but that also were, at some point, physiologically continuous portions of it, or of other plants that were themselves portions of portions of portions of it.[6]

A certain living, physical continuity thus persists in every clone-child with every parent preceding it, leading to a form of biological immortality for multicellulars. Just as was the case for unicellulars, the immortality of clonal plants, fungi, corals, lichens, and algae implies nothing about individual identity. It simply testifies to a persistence of metabolic activity and (relative) genetic stability across events of division. Critically, it is this genetic stability that, rightly or wrongly, is thought to differentiate the continuity of clonal immortality from the discontinuity that marks the life and death of sexually reproducing organisms. Nevertheless, the complication introduced by multicellulars consists in the novel presence of individual, differentiated, organized wholes, each of which may (or in certain cases may not) lead their own lives independently of their clonal offspring.

As we have seen, the presence of these part–whole relations between otherwise autonomous (or potentially autonomous) organisms leads to many complications of identification. We do not yet know how to read the meaning of these relations. If the indifference of biological or metabolic immortality toward individual identity posed problems for the worlds of unicellulars, then such difficulties will only be amplified for those of differentiated multicellulars.

There are two reasons for this amplification. The first is the evident difference between clonal and sexual forms of reproduction. Though this difference already exists in a fashion within the worlds of microorganisms—in the forms of binary fission and lateral gene transfer—it becomes more pronounced within those of macroorganisms as a result of the presence of differentiated, organized wholes. It is precisely this difference that Hayflick implicitly denies when he claims that biological immortality exists, to the extent that it exists, equally for populations of humans and microbes. Although this claim may have validity at a certain scale, it cannot be taken for granted.

The second reason for this amplification is that within the worlds of multicellulars, the limit between individual and other is often contextual. Every dividing zygote is already a clone. Those clones that remain networked with other clones of a single zygote may, as a result of this proximity, develop in such a way as to contribute to the overall success of the organism of which they will eventually form a part. Alternatively, they may go rogue, become cancerous, and revert to ancestral programs of propagation, at the expense of that whole. Those clones that happen to become cut off from their clades (both informationally and spatially), if they remain viable, may instead form new organisms. Yet in doing so, depending on their inherited adaptive strategies, they may also come to contribute to colonies or superorganisms—formations that cannot be categorized in a simple or straightforward fashion as either groups or individuals. *Dictyostelium* slime molds, quorum-sensing bacteria, and cancers are all instances of normally single-cell organisms that, under the right conditions, also betray contextual forms of collective and multicellular organismality.[7] Moreover, as Thomas and Rose point out in this chapter's epigraphs above, such organizational plasticity is all too common in the plant kingdom, where the limits between self and other are often difficult, even impossible, to identify. The point is that just as soon as one begins to ask after the limit of the organism (be it asexual or not), difficulties abound not only outside—in terms of its genetically identical and physiologically branching autonomous progeny—but also inside—in terms of its basic constitution as an individual, given the possibility either of radical morphological change within a single ontogenetic life cycle, or of highly complex, multigenerational, and unigenerational groupings. And we have not yet even begun to consider transspecies symbioses, without which no complex life could exist!

Even if we wish to accept all the abovementioned difficulties associated with identifying biological individuals qua individuals, it remains the case that the legibility of the difference between clonal propagation and indefinite aging depends on the presumptive place of the individual. In the former case, it is presumed that a structurally stable individual begets one, or multiple, new, autonomous individuals: a daughter or scion. In the latter case, it is presumed that a structurally stable individual persists as that individual. Both conceptions thus presuppose organismal individuality as a fundamental unit, even if it is only in the second case that, most of the time, the organism is believed to be extended and immortalized, in the full senses of these terms. The essential point is that the very difference between clonal propagation and (indefinite) aging becomes difficult to conceive if we abandon the notion of individuality.

By accepting that organismal individuality may always be contextual—both in terms of what it means to be an individual and in terms of the stability of any given organization, for groupings of cells, within a lifetime—we acquire the ability not only to account for a much wider range of biological phenomena but also to understand how relative morphological stability, such as mammals enjoy, represents one developmental strategy on a continuum with many others. At the same time, such an approach will also force a reassessment of the meaning and presence of death—both as a biological and a phenomenological concept—internal to these now-flexible assemblages. Simple empirical determinations of death, based on sheer material loss, can no longer be adequate to account for these life cycles. Yet the question remains: in what ought the individuality of the individual consist? Where does one organism begin, and end, within any given context? And if no single universal standard for organismality, or individuality, can be identified—because, on the one hand, scaling up from the single cell is inadequate to account for emergent features of multicellular organisms, and, on the other hand, because no single model organism exists from which to measure all the individualities of the others—what, then, would the limits of a meaningful conception of either be, for us, in our quest to live forever?

Sexual Reproduction versus Cloning

Before proceeding to investigate the various forms of indefinite aging, we need to revisit one of our working assumptions as well as its biological grounds. For the purposes of the present discussion, we have been focusing on clonal forms of growth: the grafting of a new tree, the zygote as clone. We have done so to the exclusion of sexual forms of reproduction. The reasons for this seem to go without saying. Even without recourse to genetic knowledge, the production of a child from two (or more) parents seems to signify a rupture between the generations.[8] That is, sexual reproduction is the model for intergenerational difference. It so happens that at the molecular level, this experiential difference is also, to an extent, borne out, but it is now time to ask whether the simple exclusion of sexual reproduction, and its strict opposition to clonal (parthenogenic or asexual) forms of propagation, is truly justified. To what extent does sexual reproduction represent a greater rupture than clonal propagation, such that we are (evidently) warranted in considering the biological immortality of the latter but not the former? In what consists the link that persists between clonal parent and child, supplying material, functional, and even spatial continuity between one organism and another, such that it becomes difficult to say where one ends and the other begins? In what consists this identity—or, at any rate, this apparent identity—that is not otherwise present in the lineages of individuals linked through sexual procreation?

The plant spray is a fracture of one organic unit that can grow into another such unit. One may well wish to problematize the unity of the original organism; any given apple tree, for example, can only live as an apple tree by virtue of its constant interactions with its environment—the energy it takes in from the sun, the mycelial networks that connect it to other trees, the millions of microorganisms that inhabit it and play fundamental roles in its development and growth. Moreover, the tree is itself no less a population of modular units, or metamers, which interact and compete with one another. At any given time, within the whole that we call a tree, there will be any number of such modular units, including ones that are both physiologically living and others that are physiologically dead, making it inherently multiple, differential, and nonselfsame. The "it" of the apple tree may well just be our projection on what is at best a colony of semi-independent holobiont units that happen to share a single host body and, through that proximity, are forced to divide communal resources like so many tenants in a hotel.[9] Though he had no knowledge of molecular biology—or of the existence of stem cells, let alone the DNA interior to them—this possibility, and the difficulties that follow from it, were already apparent in a nontrivial fashion to Kant. Writing some 230 years ago in the *Critique of Judgment,* Kant perceives that the individual twig's potential for grafting—the virtual possibility of its independence—is sufficient to undermine any facile notion of individual identity. A protoconcept of contextual organismality thus appears at the very moment that the no less significant notion of self-organization does:

> One part of this creature also generates itself in such a way that the preservation of the one is reciprocally dependent on the preservation of the other. An eye from the leaf of one tree grafted into the twig of another brings forth a growth of its own kind in an alien stock, and similarly a scion attached to another trunk. Hence one can regard every twig or leaf of one tree as merely grafted or inoculated into it, hence as a tree existing in itself, which only depends on the other and nourishes itself parasitically. At the same time, the leaves are certainly products of the tree, yet they preserve it in turn, for repeated defoliation would kill it, and its growth depends upon their effect on the stem.[10]

Although the terms of the debate have since changed, the problem of organismal modularity, or contextuality, remains no less difficult for today's biologists when it comes to identifying the limits of organic wholes. Some, such as Levin, perhaps still inspired by a Kantian insight, have even concluded that the seat of individuality of planarians lies not in the whole, physiologically contiguous worm but instead in each stem cell contained within, and hence in the organism's minimal regenerative unit.[11] Be that as it may, and however problematic the ascription of identity to a tree, a planarian, a metamere, or a stem cell may be, one is hard pressed to deny altogether the existence of biological wholes. To do so would be to neglect the emergent properties and unique morphologies that only exist at certain levels of integrated biological systems.

Understanding these complexities internal to organisms is critical if we are to grasp not merely the difference between sexual and clonal forms of reproduction at a physiological level but also, more importantly, the significance of this difference for questions of individuality, death, and immortality. The meaning of clonal versus sexual propagation can only be comprehended in light of the multiple and varied forms of plasticity already at work in biological wholes.

On one level, then, the question of sexual versus clonal propagation remains entirely secondary to the ontogenetic organizations through which biological individuals come to be. The pertinence of the question arises, however, when it comes time to amplify, or diminish, the gap that ought to persist between these autonomous organisms. These forms of propagation, now undergirded by knowledge of genetic mechanism, serve alternatively to reify or deny the existence of such gaps. That is, they serve either to affirm or reject the existence of generations as generations, which is to say as separate or identical. Knowledge of the genetic mechanisms separating sexual from asexual modes of propagation can thus, for example, be mobilized in the name of long-standing, problematic notions of ontological and species immortality, which presume that mere biological autonomy (such as animals and plants have been thought to enjoy) is insufficient to speak to true ontological difference (reserved for humans alone), and hence to any real distinction, or even variation, among members of a species. The "clone," with its alleged genetic basis, thus replaces the "animal" as the being whose being is marked by indifference. The opposition between clonal and sexual reproduction can thus serve to reproduce the wholly metaphysical opposition between human life and bare life. But why is this? In what consists the specificity of clonal propagation, such that it allows us to maintain an identity of mother and daughter, in contrast to the difference between the parent and child of sexual procreation?[12]

The spray, whether a shoot or a single cell preserved in vitro, seems to preserve something of the organism from which it comes, allowing it to live on in a new plant in a way that sexual reproduction does not—that, anyway, is the conceit of cloning, which itself relies on a logic of transferrable parts or portions. The life and death of sexually reproducing organisms, by contrast, appears to beget difference with each event of birth. Thus, when one clones one's dog, for example, by implanting into the nucleus of a donor egg portions of a given canine's DNA, just as with the plant spray, there is some continuity of cellular (and not just informational) material that lives on from the original in the clone. A part or portion of the first dog (one of its cultured cells, whose nucleus contains its genetic material) is preserved in the second (or third, or fourth, or fifth). The most significant difference between this process and canine (or vegetal) sexual reproduction is that in the latter case, there is no preservation of the genome. Because of the crossing over of genes that takes place in the combination of gametes from sperm and egg, the resulting zygote ends up not resembling either parent, genetically speaking.

The new zygote is thus a novel genetic entity, insofar as it combines elements from both parents' DNA. The totality of the DNA of the offspring, in other words, does not reflect the totality of either parent's DNA. Nevertheless, there is still continuity, even in this latter process—one no longer at the level of the genome but instead at the level of the individual gene. The question thus becomes what justifies or motivates the privilege allotted to one of these levels—or scales—above and beyond the others. At a certain, nontrivial level, as Walker points out, all known life today descends from the same source; it both embodies and represents the historical, material, and function survival of some of the first self-replicating organic molecules. There is anatomical, physiological, and spatiotemporal continuity between these first replicators (or, to be precise, the last universal common ancestor) and all subsequent life—that is, to the extent that there is continuity of this kind at all. If, however, we privilege today the dividing bacterium or the regenerant bud of the apple tree or planarian—rather than, say, the individual gene or even the property of reproductivity as such, which must also exist as potential immortals (that is, if they exist at all)—it is because we have come to identify in the former, but not latter, higher and more complete, perhaps more culturally and socially recognizable, degrees of continuity and similitude. In the case of the in vitro clone, it is that of the genome as a whole. In the case of the bacterium, plant cutting, or ablated planarian neoblast, it is not only the genome that persists but also an entire, living, and self-contained autopoietic system with an identical epigenetic history to the organism from which it comes—that is, up to the point at which it is divided off.[13] What is preserved, in short, is not only gene or genome but also metabolism, the surest biological evidence of vitality available to us.

The difficulty with sex is thus that it interrupts the continuity of the genome, even if it does not interrupt basic metabolic continuity, which is everywhere preserved at the minimal level of the single cell.[14] Hayflick is thus perhaps not incorrect when he points out that immortality exists equally for populations of microorganisms and for populations of humans. In both cases, there is preservation at the level of the individual gene, and perhaps that of the species. However, if one simply equates these forms of preservation, despite all the differences between them, then we could ask, why stop there? Why privilege these scales of organization rather than, say, what binds together all of life as a whole?

The difficulty with asexual reproduction, then, among multicellular organisms is that metabolism and genetics only capture a part of the whole individual. Yet the difficulty with asexual reproduction among unicellular organisms, as we have already observed, is that duplication already appears to entail the disappearance of the individual, or the original whole, which is wholly taken up by the process. Thus, while an apple tree can reproduce sexually, just as bacteria can share genetic material in conjugation, it can also simply split off a branch. Under the right conditions, that branch can regenerate into a new apple tree, virtually identical to the former, genetically and morphologically speaking.[15] Between generations of apple tree clones, there is therefore an identity of kind, one we could call a qualitative identity. This identity persists, even if any given apple tree itself has a limited life-span and a unique life. Individual trees die, but the "same" trees also live on in their genetically identical, and physiologically continuous, twins. The most famous such case is that of Newton's four-hundred-year-old Flower of Kent, which survives today in the Cambridge Botanical Gardens in front of Trinity College, as well as across the pond at the Massachusetts Institute of Technology.[16] Cases like this testify to the persistent pull of basic biological forms of continuity, even for the most scientifically minded.

12

BIOLOGICAL IMMORTALITY III

Multicellular (Organismal)

> If we did not find any replacement of cells in the higher organism, we should be induced to look upon death itself as the direct result of the division of labour among the cells, and to conclude that the specific cells of tissues have lost, as a consequence of the one-sided development of their activities, the power of unending life, which belongs to all independent primitive cells. We should argue that they could only perform their functions for a certain time, and would then die, and with them the organism whose life is dependent upon their activity. . . . But the replacement of cells is certain in many tissues (in glands, blood, etc.), so that we can never seek a satisfactory explanation in the train of reasoning indicated above.
>
> It must be admitted that we can see no reason why the power of cell-multiplication should not be unlimited, and why the organism should not therefore be endowed with everlasting life. In the same manner, from a physiological point of view, we might admit that we can see no reason why the functions of the organism should ever cease.
>
> —August Weismann, "The Duration of Life"

Cloning is not the only path through which multicellular organisms maintain metabolic continuity and thereby stave off death for indefinite periods of time while undergoing relatively little genetic variation. Several other multicellular organisms defy death simply by not senescing. Specimens of the Great Basin bristlecone pine are known to have survived for at least five millennia.[1] Thus, whether or not such pines reproduce, individual members of the species live on, in their own bark, for extreme—if not simply indefinite—lengths of time. The reason they can do this, it is believed, is because they are able to grow, regenerate, and reproduce, all without bearing the cellular signs of degeneration commonly associated with aging. As authors of a 2001 publication in *Experimental Gerontology* put it, biologically speaking, "the concept of senescence does not apply to these trees."[2]

Of course, individual members of the species change. They do not simply stay the same. They grow and develop. They suffer environmental incursions. Moreover, they frequently deplete their soil of nutrients, making their survival more difficult. They also reproduce. But insofar as the biological concept of aging requires that an organism betray imprints, or biomarkers, differentiating old from young in terms of relative senescence, to that extent, such a standard simply does not apply to these trees, which express none. Thus, according to scientists, there may be no upper limit on their life-spans. They may be naturally immortal. As we shall see, this form of biological immortality implies an entirely different relationship to organismal individuality than did that involved in clonal propagation. Because it now supports the perpetuation of an organizational totality (not necessarily that of the individual cell but instead of the structures composed and the functions performed by collections of cells), this form of biological immortality seems to open the door for an indefinite prolongation of the lives of individual organisms.

According to Lanner and Conner, the key to bristlecone cellular youth lies in their meristems: the tissues that are found under the surface of their roots, trunk, and limbs. Meristems are groupings of "totipotent cells"—a variety of plant stem cells—capable of producing any other cell types needed for the tree's growth.[3] Because the tree's bark insulates the meristems from most of the damage suffered on the outside surface of the tree, the meristems can secure an exceptionally high fidelity during cellular reproduction, effectively minimizing mutations and keeping the trees physiologically young.

According to David Sinclair, Harvard biologist and author of *Lifespan: Why We Age—and Why We Don't Have To,* the key to human eternal youth lies in something just like what the bristlecone pine manages to do: minimize epigenetic noise. For Sinclair, the effects of aging (or biomarkers of senescence) can be explained through the cumulative loss of information that occurs over a lifetime's worth of cellular reproduction, damage, and repair. Aging, rather than being something genetic or preprogrammed, describes the consequences of epigenetic loss on the genome as more and more cells inevitably forget their original bodily functions, as a result neglecting essential elements of what Sinclair calls the survival circuit.[4] Yet far from simply inevitable, because aging is best understood as a function of the epigenome, it is for Sinclair ultimately the most malleable of physiological realities, a disease that can and should be eliminated. If one can minimize the "inherent instability of the genome"—as the bristlecone pine and planarians evidently have evolved to do—then one can minimize aging.[5]

The bristlecone pine grows, but it does not age. It survives the passage of time, but it shows no biomarkers of senescence. It may be helpful at this point to introduce a terminological distinction used by botanist Howard Thomas. Whereas Sinclair and others frequently refer to aging and senescence interchangeably, for Thomas, there is a slight but significant difference between these terms. Senescence, says Thomas, refers to a "phase of development" following the completion of growth, which may or may not result in death.[6] Aging, on the other hand, emphasizes time-based processes and "is not simply another name for declining viability."[7] Rather, aging refers to the bodily experiences of change over time, whether or not these experiences result in deterioration: "ageing is the *time-stress response*."[8] Thus, while growth, aging, and senescence can coincide, what the work of Thomas and Sinclair shows is that they need not; a plant or an animal can grow from presexuality to sexuality. Insofar as this sexual development occurs in time, it will correspond to a certain amount of aging of the organism. Nevertheless, I can age without growth, and neither aging nor growing need correspond to senescing, which—if we follow Sinclair's logic but use Thomas's terminology—would refer specifically to the epigenetic noise that accrues in our cells but that does so at a nonlinear rate with respect to the sheer passage of time (aging) or to the cellular development of organisms (growth).

To return to the bristlecone pine, we could say that although it ages, it does not senesce. In this respect, we could contrast its form of immortality with that of Tithonis, who, at Eos's request, is given the gift of immortality by Zeus. On that day, Tithonis becomes immortal—without, however, being given the gift of nonsenescence. Although Zeus permits Tithonis to not die, he nevertheless (evidently) does nothing about the accrual of epigenetic noise that occurs through entirely natural processes of cellular reproduction and repair. In other words, while Tithonis cannot die, he does age, and he does senesce. In this sense, the pine may yet have a trunk up on him.

For Sinclair and many others, the key to human immortality lies in a model like that provided by the pine: one may grow and age, but in so doing, one should refuse to senesce, and in this way refuse to die.[9] So long as I can be confident that I am myself, immortalization is merely a matter of extending the duration of my (given) vitality. Problems arise with this model only if I begin to question this identity, or the feasibility (technically speaking) of such a biophysical transformation for *Homo sapiens*. As we have already seen, questions of identity do in fact emerge constantly in the world of plants; there are both living and nonliving metameres within any given pine. Can any given metamere persist indefinitely? And does it even matter if the tree as a whole can persist? However we come down on these questions for the tree, the question of identity carries a particular valence when it comes to the worlds—and dreams—of humans, which have no simple equivalents in plants. We will examine the nature of this difference more closely later, when we return to the problem of metabolism and its relationship to phenomenological identity.

If we now move from the plant to the animal kingdom, we find yet other organisms that betray similar patterns of nonsenescence to that of the pine within the phylum Cnidaria. According to Daniel E. Martínez, over the course of a four-year observation of hydra cohorts, individual hydras showed no signs of senescing. The hydra, he writes, "may have indeed escaped senescence and may be potentially immortal."[10] Unlike the bristlecone pine, however, *Hydra vulgaris* does not do this through the protection of meristems and efficient cell repair. Instead, it shows a capacity for "constant renewal of its body," which consists of

> only about 20 cell types that are organized into three cell lineages: two epithelia and one interstitial. Each lineages *[sic]* consists of a population of stem cells with indefinite self-renewal capacity, and several differentiation products. . . . A striking characteristic of hydra is its tissue dynamics. The epithelial cells

> of the body column are continuously in the mitotic cycle. This continuous production of epithelial cells . . . is balanced by a loss of cells: 85% of the epithelial cells are incorporated into developing buds while the rest are sloughed off. . . . A result of these dynamics is that epithelial cells are constantly displaced either apically onto the head, or basally into developing buds, or onto the foot.[11]

At the cellular level, the hydra is in a state of constant flux that is proportionally much greater than that of other metazoans. No individual cell persists for very long because of the double effect of, first, the high rate of mitosis and, second, the displacement of "nondividing differentiated cells," which are lost from the body column at its various extremities within twenty days.[12] The key to the renewal and regeneration of hydras, however, and what ultimately separates their growth and repair from that of other multicellulars, is the pluripotency, or plasticity, of their interstitial and epithelial cells, which consist largely of stem cells. Like the meristems of bristlecone pines or the neoblasts of planarians, the stem cells of the hydra possess the capacity to dedifferentiate and be reprogrammed. Whereas, in adult humans, stem cells are found only in small numbers—for example in the brain, gut, and blood—in the hydra, they are numerous enough to allow for sixty full replacements of its body within a four-year time span.[13] As Martínez puts it, in the hydra "cells are either constantly renewing by cell division, or they are lost from the animal in a relatively short period of time. This capacity for constant renewal is the main reason behind the claims that hydra is potentially immortal."[14]

What bristlecone pines and hydras become emblems for is thus the relative stability of self-sustaining, multicellular organization. These organisms can grow and age, yet without senescing. The jellyfish *Turritopsis dohrnii* (as well as *Turritopsis nutricula*) is another famous case of a cnidarian with death-defying properties—but now with a twist. Bristlecone pines can survive indefinitely on the condition that they do not senesce, but this jellyfish species does so by senescing and desenescing in turn.[15] What distinguishes *T. dohrnii* is its singular ability not only to evolve from polyp to jellyfish but also to devolve from jellyfish to polyp, and back again, through transdifferentiation. Transdifferentiation allows developed cell types to convert back into undifferentiated polyp cells. This process of "ontogeny reversal" resets their biological clocks, allowing for the theoretical possibility of indefinite cycles of growth and rejuvenation.[16]

Between the jellyfish and the hydra, there is thus a difference of degrees: both possess an incredible capacity for renewal. Renewal is possible because they turn back their biological clocks by turning over their cells. Yet in the latter case, this is an ongoing process. Hydras age, but they do not show significant biomarkers of senescence; nor does their cellular turnover correspond with any significant morphological change (that is, after reaching sexual maturity). For *T. dohrnii,* by contrast, one can indeed distinguish between distinct ontogenetic developmental stages.[17] A jellyfish will thus pass from polyp to medusa and back again. There is both development and senescence, but there is also reverse development, and with it rejuvenation. "The medusa is distinguished from the polyp, not only by a different shape and anatomical organization, but also by a completely different set of somatic cells in the umbrella."[18] This is what is miraculous about the jellyfish.[19] It can reverse the order of its development even after reaching sexual maturity, making it "the first metazoan known to revert to a colonial, juvenile morph after having achieved sexual maturity in a solitary stage."[20] A study of "Reverse Development in Cnidaria" compares this process to that of a "butterfly . . . able to revert to its caterpillar stage."[21] The difference here is that in the case of the medusa, back-transformation entails the formation of an entire polyp colony of medusa buds, from which multiple genetically identical medusae are birthed in turn: "For each medusa undergoing reverse ontogeny, a new colony can potentially originate, which in turn will produce new copies of the original medusa, leading to an exponential increase in the total number of clonemate offspring."[22] As with bacteria, it is difficult to identify an empirical event of death in this process. Nevertheless, the transformations that medusae undergo, and the clonal duplications that result, should force us to reconsider the nature of the continuity at play here, as well as of the identity or selfhood endemic to each organism as a whole. To repeat what is becoming a common refrain: while questions of continuity and selfhood appear to matter little to the cnidarians, they nevertheless matter quite a lot to us, and this fact should give us food for thought.

13

METABOLISM AND THE MEANING OF LIFE

> Metabolism can very well be considered as the defining quality of life: every living being has it, no nonliving being has it.
>
> —Hans Jonas, *Organismus und Freiheit*

We have observed several pathways through which organic life-forms apparently manage to defy the so-called inevitability of death. They do so by way of specific biological mechanisms. These mechanisms include binary fission, vegetative propagation, cellular regeneration, and reverse ontogeny. It was observed, moreover, that in each case, the recognition of such mechanisms not only as technologies of hereditary survival but also possibly as means of individual immortalization required prior reference to the lives of the individuals thereby extended. Only on this basis, only because there is something that lives—an "I" or perhaps a "we"—could one in turn identify these solutions as solutions to the problem of death. As we shall see in Part IV, it is also on the possibility of identifying organic solutions to death such as these that depend transhuman and posthuman speculations on inorganic ones, such as brain uploading, human cloning, and antisenescence technologies, as well as various bridge technologies, such as cryogenics. The prostheses of immortalization of late capitalist speculation are underpinned by those of empirically observable organic nature, made possible by advances in the fields of physiology, neuroscience, and microbiology. It is this basis, in the natural sciences, that is believed to separate these efforts from their religious, literary, and alchemical antecedents.

Thus, having established in what these basic mechanisms of indefinite biological survival consist—as well as in what their drawbacks lie—we must now turn back to the nature of both this life and the individual that these mechanisms variously presuppose. It is only because there is already some life, as well as some form of identity, or ipseity, inherent to or emergent in it that one can imagine this life's projection, either into the future or within the present of the present. Only because of this prior determination of life can one presume forms of continuity between self and other, or what is here even more fundamental: between self and self. Such determinations, it should by now be clear, are always fraught. Yet just as soon as one wishes to understand the meaning of the forms of continuity that persist in the biological world—be they those of mere metabolism, of genetic identity, of articulated, organismal individuality, or of personhood—one cannot but contend with them.

The purpose of each of the previous discussions, in shedding light on these difficulties, was not to decide on these matters. It was rather to highlight how decisions on each of these borders remain undecidable, and not only for us, but also for any future superintelligence or any past god. Each decision, border, or limit is less a matter of an either/or, firmly sedimented in its ground, than it is one of an economy of differences, each of which is recognizable only at certain, specific scales. To recognize and reveal the performative fictions at the heart of the living cannot, therefore, mean simply to discount the validity of the existence of life, or afterlife. But neither can it be to blindly accept these fictions and the hegemonies of blood, race, nation, species, and culture that may follow in their train.

To this point, we have negotiated with various economies of sameness and difference, across and within many fantastic animal, plant, and bacterial bodies. The final resting place for each of these discussions lies, however, in the problem of metabolism, to which we now turn. According to a significant tradition of biophilosophical thinking, metabolism is considered to be the process whereby physicochemical relations come to signify as value within the performative closure of their operational systems. In other words, metabolism names the point of contact between mechanical nature and symbolic experience. On the one hand, from the materialist point of view, it is the process whereby information emerges as a causally efficacious property of physical systems. On the other hand, from the phenomenological perspective, metabolism is recognized to be the originary site of the emergence of the difference between the material and the symbolic. That is, it is taken to be the event (or process) whereby anything like difference can arise in the first place, because it entails the worlding of the world in care. Thus, although it has different senses in each of these discourses, metabolism nevertheless names for both something like the essence of the living: that through which the presence of life presences. If we wish to understand how individuality, ipseity, or selfhood emerge, such that they can become possible targets for life extension, then understanding the meaning of metabolism evidently holds one key.

At stake in the following discussion will therefore not simply be the nature of organic, physiochemical processes but, more precisely, the operations by which these processes can become meaningful, for an organism, in the becoming-organized of matter—by which they can become meaningful for an organism that, truth be told, does not precede these processes but instead arises through the very difference that they make possible, in the affirmation of that difference. Thus, while metabolism is not simply continuous with individuality—or, rather, while metabolism clearly cannot speak to all of the variegated forms of individuation, as we have already observed—it is nevertheless both a necessary and sufficient condition for individuality's minimal articulation. Where there is metabolism, we could say, there is emergence of world. And so, having previously explored the various ways organisms extend themselves, thereby bringing into question the necessity of death as a physiological phenomenon, here we will turn at last to the basic fundament on which all these other extensions evidently rest: the basic constitution of the living present, as the birth of world, in care, from the differential operation of metabolic processes. In this way, our investigation goes beyond the sense and structure of the seemingly rational projections of biologists when they speak of indefinite biological immortality, reaching all the way back to the structural and biological origins of this immortalist projection itself. It is no longer, in other words, merely a question of the objective discourse of the life sciences on the biological field, but a question of the nature of the simplest and most basal projections of organisms themselves (be they human, animal, bacterial, fungal, or vegetal) as they homeostatically orient the conditions for their own survival.

All pretensions to life extension—natural or radical; organic or inorganic; or biological, technological, or informational—so long as they presume to extend a feeling being, must refer back to these basic metabolic reactions, which supply the ground for all life as we know it. If, as I shall argue, all such extensionist efforts do in fact refer to this foundation, then what that in turn means is that the latent potential for prosthetic supplementation—to which every form of extension testifies—may already be contained in seed within these basic metabolic operations. Prosthetic immortalization can best be understood as a consequence of these rudimentary processes, which are endemic to survival. What metabolism (along with the closely associated concept of homeostasis) discloses then, is the essentially prosthetic nature of autopoietic living systems. They do so by revealing the performative fiction at the heart of biological autonomy.

Metabolism, Self-Organization, Autopoiesis

Though the term "metabolism" first enters English in the seventeenth century, then begins to take on a specialized, biochemical sense in the nineteenth century, its importance as a key figure for the contemporary conception of living systems as self-organized can be traced to the writings of Hans Jonas, a twentieth-century German philosopher and student of Heidegger. The distinction of Jonas's work consists in taking what had been conceived as a merely mechanical process, in the observable transfer of matter and energy between an organism and its environment, and in perceiving how this process is implicated in the emergence of its world, and hence in its becoming an individual. Andreas Weber and Francisco Varela note that what Jonas's work offers is a "re-enchantment of metabolism."[1] This "re-enchantment" concerns specifically the establishment of the link between physicochemical reactions and organismal ontology.

Kant, of course, was the first to describe the living as a matter of "self-organization."[2] Nevertheless, the novelty of Jonas within this tradition is to turn the problem of natural teleology, still suspended by Kant within an "as if," into an emergent feature of self-organization. Thus, it is no longer a question for Jonas of knowing whether the purposiveness apparently at work in living systems is actually or only seemingly one of their intrinsic features. For Jonas, metabolism (as a distinctly biological form of self-organization) gives birth to an interior perspective, from which point alone it becomes possible to speak of autonomy and intrinsic teleology. Metabolism, in sum, performs teleology. In the words of Weber and Varela, "Instead of viewing meaning or value as artifacts or illusions, both [philosophy of biology and theories of autopoiesis culminating in the 1970s] agree on a new understanding of a form of *immanent teleology* as truly biological features, inevitably intertwined with the self-establishment of an identity which is the living process."[3] Thus, for Weber and Varela, it is Jonas who develops an interpretation of Kantian self-organization that does not reduce the living to the binary of mechanism and teleonomy, but instead unpacks the possibility of " 'intrinsic teleology' arising from biological autonomy and biological individuality."[4]

Jonas's insights into metabolism, which he first developed in the 1960s, would in subsequent years become an integral element of the thinking of autopoiesis in work by Maturana, Varela, and Uribe, as well as a central figure for the concept of biosemiotics in Hoffmeyer.[5] Today, it remains a critical component of the thinking of biological autonomy (for example in Moreno and Mossio), of enactivism (in Evans and Di Paolo), and an implicit, even if at times unavowed, antecedent of the theory of homeostasis (in Damasio).[6] Metabolism, in sum, in the wake of Jonas, has come to name the dynamical process whereby life is performed and through which, as a consequence, any decision on the survival of what survives must pass.

I will first explore how contemporary theorists take up metabolism in their own work before returning to Jonas's writings to see how it contributes to the emergence of individuality there. Just as soon as organismal individuality is revealed to be the consequence of metabolic processes, ipseity cannot but be understood as the first prosthetic production of life. Life itself, however, in light of this prosthetic performance, can no longer be said to precede this act.

Lynn Margulis, originator of the endosymbiosis theory of organogenesis, explains the sense of metabolism and its relation to autopoiesis:

> Metabolism is the name given to this incessant buildup and breakdown of subvisible components—that is, to the chemical activities of living systems. If physiology is the study of the functions of living organisms and their parts, then metabolism is the chemical manifestation of those functions. Metabolism can be defined as the sum of the enzyme-mediated network of

> chemical and energetic transformations of living systems. It is more easily understood as the incessant movements of matter that occur all the time in living systems and that cease when the system dies. Autopoietic systems metabolize, whereas nonautopoietic systems do not. Proteins, viruses, plasmids, and genes are all components of live material. When contained within the boundaries of animal, plant, or other cells, they may be required to sustain cells or organisms and their autopoietic behavior; yet proteins, viruses, plasmids, and genes, intrinsically incapable of metabolism, are never autopoietic in isolation. Metabolism includes gas and liquid exchange (breathing, eating, and excreting, for instance); it is the detectable manifestation of autopoiesis. Autopoiesis determines physiology and hence is the imperative of all live matter. Autopoietic entities, that is, all live beings, must metabolize. These material exchanges are the *sine qua non* of the autopoietic system, whatever its identity.[7]

Metabolism, from the Greek μεταβολή (throw across), originally bore the sense of change before it took on the sense of commercial exchange.[8] Within the contemporary biological context, it names the steady exchange of matter and energy perpetuated by all (known) living organisms. What marks this exchange for such organisms (in contrast to nonliving systems) is that this cycle is not only perpetuated by them but is also a requirement for their very self-maintenance as organisms.[9] That is, without metabolism working in the service of autopoiesis, not only would living beings cease to be able to act, but also their very informational and physical bases—through which action first becomes realizable in terms of concrete possibilities—would themselves decompose.

The kind of self-maintenance at work in living systems, if it is truly to carry the sort of explanatory weight for life that proponents of autopoiesis and metabolism wish it to, should therefore serve to distinguish these systems both from organized, teleologically oriented technical systems, such as internal combustion engines and missile guidance systems, and from open, far from equilibrium, but (evidently) nonliving dissipative systems, such as tornadoes and Bénard cells. The thinking behind such divisions goes as follows. On the one hand, while the former may possess elaborate feedback loops that allow them to sense and respond to stimuli for the purposes of self-regulation, it is only for the living that both the material substrate and the organization of the system depend, for their very stability, on the maintenance of their metabolic activity, without which the system's component parts would break down. Mechanical systems, by contrast, tend to perdure whether or not they are in use. On the other hand, while naturally occurring Bénard cells and tornadoes may be self-reproducing, integrating foreign matter and energy into their internal organization, they do so by way of universal physicochemical laws. In other words, both Bénard cells and tornadoes arise without forming second-order networks of constraints. The metabolic exchanges of living organisms, by contrast, are mediated by top-down processes that are historically determined by dynamics arising internal to these organisms, which these organisms play a foundational role in establishing. Living systems, in sum, in distinction to spontaneous, far-from-equilibrium dissipative systems, establish their own boundary conditions.[10]

Thus, whether or not future technical systems will one day be able to reproduce the complex interdependencies of living systems—a question that, while compelling and certainly not excluded by the present discussion, nevertheless is not relevant for it—it is precisely these complex interdependencies that are thought to mark the specificity of the living as composing autonomous, rather than heteronomous, systems.[11] In other words, the reason that living systems have evolved to be the way they are, and the reason that they bear the distinctive attributes they do, such as high degrees of plasticity, adaptability, open-ended evolution, and semantically meaningful relationships with their environments, are their unique kinds of dependency, which are based in evolved forms of metabolism, and which forms can therefore be said to be both their cause and their effect.

In the *Critique of Judgment,* Kant attempts to explain the purposiveness of self-organized beings:

> A thing exists as a natural end *if it is cause and effect of itself* (although in a twofold sense).[12]

> In such a product of nature each part is conceived as if it exists only *through* all the others, thus as if existing *for the sake of*

> *the others* and *on account of* the whole, i.e., as an instrument (organ), which is, however, not sufficient (for it could also be an instrument of art, and thus represented as possible at all only as an end); rather it must be thought of as an organ that *produces* the other parts (consequently each produces the others reciprocally), which cannot be the case in any instrument of art, but only of nature, which provides all the matter for instruments (even those of art): only then and on that account can such a product, as an *organized* and *self-organizing* being, be called a *natural end*.[13]

A structure (or artifact), such as a heat-seeking missile, may be both teleological and organized. Insofar as it incorporates negative feedback in order to modify its behavior, it is teleological. Insofar as its component parts are designed to reciprocally interact, it is organized. Yet to be fully self-organizing, according to Kant, a natural end must be both cause and effect of itself. This means that it must additionally account for its very existence. In the *Critique,* Kant identifies three ways in which this may occur: first, in the generation of another organism, through species; second, in the generation of the individual, through growth; and third, in the preservation of the individual, through the reciprocal interdependence of its parts.[14] Autopoietic systems are examples of such self-organizing structures. In particular, they are the subset of self-organizing structures that depend in fundamental fashion on metabolism for the achievement of their self-organization.

Theories of autopoiesis, generally speaking, attempt to take account of the forms of self-organization endemic to living systems; they attempt to ask how the architectures of self-organization affect the ways these systems function and develop. They attempt, in light of these forms of interdependency, to render explicit the essential role of both physical and symbolic boundaries to basic metabolic operations. Such boundaries, as we shall see, are both a necessary feature of organisms, without which metabolism could not function, and a product of the very metabolic activity that they make possible. Autopoiesis therefore includes metabolism but adds to its physicochemical interpretation two critical things: first, the understanding that it is through their metabolic behavior that organisms perpetuate themselves as systems capable of metabolism (final causality), and second, the appreciation that a condition of such constraint closure is the delimitation of a boundary between inside and outside (effect of individuation). In this way, Kant's understanding of purposiveness is brought into the organism and made into an emergent property of its life-sustaining processes:

> Autopoiesis proposes an understanding of the radical transition to the existence of an individual, a relation of an organism with it-self, and the origin of "concern" based on its ongoing self-produced identity. . . . Thus autopoiesis is a singularity among self-organizing concepts in that it is on the one hand close to strictly empirical grounds, yet provides the decisive entry point into the origin of individuality and identity, connecting it, through multiple mediation with human lived body and experience, into the phenomenological realm.[15]

The theory of autopoiesis was first developed by Chilean biologists Humberto Maturana and Francisco Varela in the early 1970s.[16] Its clearest expression is to be found in the individual cell (the smallest autopoietic unit, they argue). For the cell, metabolic activity can be carried out only by virtue of semipermeable barriers lying between it and its surroundings. These barriers are established by the cell wall. The cell wall, however, is itself produced and maintained only through the operations of the metabolic activity that it permits to function. Thus, from this point of view, what we call life is marked from the beginning by the implication of its survival as a biological individual in the success of its metabolic operations. Life can only be *as* self-organizing—or, at the very least, what we call or today recognize as life is marked by this complication. If metabolism names the biochemical exchange ubiquitous within all known living organisms, then autopoiesis emphasizes the further element, necessarily at work in all metabolically active systems, but not necessarily previously recognized in them—namely that such systems, in and through their metabolic activity, also constitute frames of reference for the organism, and it is only through these frames of reference that the activity of metabolism may in fact be maintained. The emergence of the autopoietic unit is, for Varela and Maturana, a thoroughly performative one. No being precedes the self of autopoiesis. Rather, autopoiesis announces the birth of self as spatial compartmentalization, which is then taken up in a process of semantic signification or self-referential information. To abuse the oft-cited maxim of Bateson, we could say that autopoiesis allows for a difference to make a difference.

The theory of autopoiesis thus develops the implications of metabolic processes within life, bringing Jonas's existential analysis in line with the empirical reality of biological organization. Returning to the minimal autopoietic system, the cell, it would be impossible to conceive of such a coordinated series of metabolic activities without there being some mechanism for feedback whereby the cell can register and respond to changes both internal and external to itself—which is to say, register and respond to changes in relation to its own effects of compartmentalization.[17] Indeed, in order for a cell to emerge in the first place, some mechanism of cellular feedback and regulation is required. Autopoiesis thus comprises not only metabolism and compartmentalization but also cognition; that is, the processes of self-sensing, through which a self is performed, and not only as a materially bounded structure separated from the world through a cell wall but also as a self-signified ipseity (even if an unconscious and unminded one) through which the empirical, material existence of the cell wall, or compartmentalization, can be sustained. As soon as there is an autopoietic system, there is cognition, which is to say meaningful information whereby the givens of the physical world are sensed *as* something, as well as sensed in relation to a self whose very being or survival is implicated by them. It is for this reason that metabolism, by way of autopoiesis, always entails world, and can perhaps eventually lead to something like consciousness.[18]

Critical to the concept of autopoiesis—and here one can see both its inheritance and its modification of Heideggerian phenomenology—is that the production of the individual, as the effect of a recursive act of compartmentalization, always implies a necessary relation to physiological (and ontological) fragility. It is only through the establishment of a boundary, which is the originary act through which a difference between self and other becomes intelligible, that there can be what we call autonomy.[19] However, the autonomy of the individual is for this reason grounded on the very thing that threatens it. The implication—or enfolding—of fragility, dependence, vulnerability, and so on of the autopoietic system (or self) within the sense-data produced by and for it, therefore announces at one and the same time the birth of self, its threat of dissipation, and its sense of need. We might even say, for this reason, that these three figures name the same phenomenon.

With the subject of need, we have arrived, at last, at the work of Jonas, for whom need is constitutive of the organism. Jonas's project consists of developing what he calls an "'existential' interpretation of biological facts" by taking phenomenological and biological insight together.[20] In this way, he ventures to close the gap between human and nonhuman forms of life through an account of their minimal existential properties. As he puts it at the beginning of *The Phenomenon of Life,*

> The following investigations seek to break through the anthropocentric confines of idealist and existentialist philosophy as well as through the materialist confines in natural science. In the mystery of the living body both poles are in fact integrated. The great contradictions which man discovers in himself—freedom and necessity, autonomy and dependence, self and world, relation and isolation, creativity and mortality—have their rudimentary traces in even the most primitive forms of life, each precariously balanced between being and not-being, and each already endowed with an internal horizon of "transcendence."[21]

Thus, by tracing higher-order, anthropological phenomena back to their biological bases, Jonas seeks to locate a common ground for understanding all forms of life. For Jonas, the most basic and fundamental biological ground is to be found in need, which marks the living, from the simplest to the most complex life-forms. Need is the expression of the "freedom" of the organism, which is always a needful freedom—a need, in sum, to exercise its freedom in order to remain free.

Jonas's work offers one of the clearest expressions of the deep interconnections between phenomenological experience and biological structure. Similar to ideas subsequently popularized by Damasio, for Jonas there can be no life without the dual implications of organization and experience, or, as Canguilhem terms it, *le vivant et le vécu,* "the living and the lived."[22] Life is, by necessity, both of these sides, and unimaginable without one or the other. It is because of this interconnection that notions of cognition can be put forward that account both for the simplest unicellular and most complex multicellular life-forms.

For our purposes, what is critical is that Jonas's reflections on the biological foundations of individuality provide a way of understanding the emergence of selfhood as endemic to life as such. Once we grasp Jonas's conception of this emergence as the performance of life itself, we can understand not only why what Derrida calls the phantasm may remain inexorable, now from the perspective of biological structure, but also why the simplest as well as most complex forms of ipseity are constitutively marked by the prosthetic, the differential, and the supplemental. Metabolism, as the prosthetic basis of biological individuality, will dictate every subsequent supplement as well as the possibilities of identification, both internal and external to this border, that is not, and has never been, *one*. It is at this point that we must ask whether the empiricophilosophical distinction between mortality and immortality, or more precisely between the prosthetic projection of self as mortal and immortal, can any longer bear any weight, if the minimal articulation of individuality of the single cell already implies a fantasy of absolute difference between self and other, inside and outside, living and dead, freedom and necessity. What if life, as originary, autopoietic self-immunization, constructs itself as the prosthesis of its own immortalization? What if life, being naturally mortal, must as a consequence posit itself as artifactually immortal?

One final word before proceeding with my reading of Jonas. Both metabolism and the autopoietic tradition it has birthed have rightly been criticized by ecologically oriented biologists and philosophers, such as Donna J. Haraway and Scott Gilbert, for their emphasis on autonomy—that is, for their focus on self-production, which risks neglecting the empirical reality of collectivities and nonindividualities. It is well known today that every human body contains trillions of nonhuman microorganisms. Lichens are symbionts, and animate life itself—insofar as we can speak of such an "itself"—is a planetary phenomenon, and one marked through and through by the inanimate.[23] As Haraway puts it,

> Yoking together all the way down is what sym-bio-genesis means. The shape and temporality of life on earth are more like a liquid-crystal consortium folding on itself again and again than a well-branched tree. Ordinary identities emerge and are rightly cherished, but they remain always a relational

> web opening to non-Euclidean pasts, presents, and futures. The ordinary is a multipartner mud dance issuing from and in entangled species. It is turtles all the way down; the partners do not preexist their constitutive intra-action at every folded layer of time and space. These are the contagions and infections that wound the primary narcissism of those who still dream of human exceptionalism. These are also the cobblings together that give meaning to the "becoming with" of companion species in naturecultures. . . . Gilbert stresses that nothing makes itself in the biological world, but rather reciprocal induction within and between always-in-process critters ramifies through space and time on both large and small scales in cascades of inter- and intra-action. In embryology, Gilbert calls this "interspecies epigenesis." . . . In my view, Margulis and Sagan's symbiogenesis is not really compatible with their theory of autopoiesis, and the alternative is not an additive mechanistic theory but a going even more deeply into differentiation.[24]

These important interventions could be supplemented from the developmental and evolutionary sides of biological inquiry. Even quantitative approaches to organismality, based in relative degrees of observable conflict and cooperation among divergent cell lines, readily admit that there is no sufficient definition of *the* organism. Nor is this simply a failing of current techniques and methodologies. Rather, it is a recognition of the complexity at work at every level of biological organization, as bacterial biofilms, insect superorganisms, cooperative tumors, and morphologically complex amoeba show.[25] The difficulty of identifying "an" organism, or "a" self, thus concerns not only the extreme, eye-opening cases of slime molds and bee colonies but also less traditionally context-dependent organisms, such as humans—which are holobionts—and unicellular bacteria, many of which can form multicellular entities, thereby demonstrating that even the simplest, single-cell, prokaryotic forms maintain their autonomy not as an absolute right or privilege but rather always already within ecological and evolutionary adaptive contexts.[26] Individuality, in brief, is not only conceptually suspect evolutionarily and phylogenetically, but also ecologically and ontogenetically, from the simplest to the most complex life-forms.

Nevertheless, in probing how organisms survive (be it mortally or immortally), our question is not ontological in nature. We are not asking whether there is such a thing as identity, self, or individuality. Instead, we are asking, in what way are effects of identity, self, or individuality performed by structures both biological and phenomenological, both material and informational? This distinction is absolutely critical, and it is what allows us to study the appearance (or phantasm) of identity as a (causally efficacious) phenomenon in its own right, even as we acknowledge that there has never been, and never will be, an individual, be it conceived on the model of the singular or the collective.[27] Autopoiesis, self-organization, and metabolism are absolutely vital ways of conceiving of how biological and phenomenological structures coemerge and coimplicate, allowing for the appearance of cognition, organism, self, life, death, and, of course, consciousness. Nor is the (very real) appearance of any of these phenomena in itself sufficient to silence debates concerning what matters when it comes to the individual, as I explore at greater length in Part IV. Be that as it may, it is only once we have accepted the deep continuities that persist between all known forms of life that we can then, in turn, adequately account for that which divides them. Thus, while I agree with Haraway and Gilbert and share their concerns, the importance of metabolism and autopoiesis is, precisely, to account for the emergence of the appearance of individuals, even and especially if one has never existed.[28] Our question might therefore be rephrased as: how does the problem of metabolism entail that of immortality (and indefinite life)? How does the mutual possibility and impossibility of individuality, within life, codetermine the possibility and impossibility of indefinite life, as a problem we simply cannot shake?

Metabolism and Individuality in Jonas

Though Jonas is extremely attentive to evolutionary development and its consequences for the organism-specific emergence of properties—such as, for example, the intimate connection among motility, perception, and emotion[29]—his most profound insights concern the common nature of basic organismal characteristics, which follow, for him, from the universally shared activity of metabolism. As he explores in his 1968 essay "Biological Foundations of Individuality," the single most significant consequence of this activity is individuality itself. In what follows, we will see what he means by individuality; we will also see how its minimal articulation, already in the single cell, affects all subsequent, higher-order manifestations of this principle. The individuality of basic metabolic activity that Jonas identifies already implies the metaphysical confusion between mortality and immortality that the philosophical, literary, biological, and religious discourses we have analyzed also encounter. As such, Jonas's account can be read as offering a prosthetic account of life through which the phantasm of ipseity emerges as a basic consequence of self-organizational principles. In his words: "My contention will be that, on the evidence penetrable to us, the proposed conditions are exhibited by organisms and organisms only, but in some measure by *all* organisms: that they are integral to organic being as such; and that therefore the realm of individuality, in all its grades, is coextensive with the biological realm as a whole."[30]

In setting out to establish the biological foundations of individuality, Jonas's primary target is the materialist reduction of the same. On the one hand, according to a long line of thinking, visible from Aristotle up through Aquinas and Schopenhauer, a "thing is individual, not in virtue of what it is, but in consequence of where and when it is."[31] On the other hand, even for Duns Scotus and Leibniz, for whom each thing has its essence "in its total autonomous history," such a determination still remained too "immanent, self-contained and absolute," making individuality "a universal possession of all concrete entities, coextensive with particular existence as such."[32] In distinction to such approaches, then, and in line with Heideggerian modes of analysis, Jonas's project is to identity individuality from within as a self-produced attribute of individuals themselves: "I wish here to take a stand 'within,' asking what it is to *be* an individual—for the individual itself; and I shall try to expound a concept which makes individuality the prerogative of a specific mode of existence and therefore the peculiarity of beings endowed with that mode of existing."[33] If we start with individuals, Jonas thus implicitly asks, how does this shift in focus allow us to reconceive the nature of individuality itself? Well, he seems to respond, rather than an external attribute, individuality becomes the effect of a process, but a process now endemic to the individuals themselves. Individuals, we could say, are those entities whose being is a matter of concern for themselves; they are those entities who are concerned with their own being. However, because their being is itself a consequence of their self-concern—which concern is realized in metabolism—their individuality both as discrete, spatially distinct (or compartmentalized) organisms and as self-concerned beings, capable of distinguishing between self and other, is a complex, processual effect:

> Only those entities are individuals whose being is their own doing (and thus, in a sense, their task): entities, in other words, that are delivered up to their being for their being, so that their being is committed to them, and they are committed to keeping up this being by ever renewed acts of it. Entities, therefore, which in their being are exposed to the alternative of not-being as potentially imminent, and achieve being in answer to this constant imminence; entities, therefore, that are temporal in their innermost nature, that have being only by everbecoming, with each moment posing a new issue in their history; whose *identity* over time is thus, not the inert one of a permanent substratum, but the self-created one of continuous performance; entities, finally, whose *difference* from the *other,* from the rest of things, is not adventitious and indifferent to them, but a dynamic attribute of their being, in that the tension of this difference is the very medium of each one's maintaining itself in its selfhood by standing off the other and communing with it at the same time.[34]

Why, therefore, must metabolism find itself at the center of this activity? And if identity, individuality, and selfhood are to become general characteristics of life, how does this bond affect the nature of life's subsequent self-projections and extensions?

Metabolism holds the key to understanding the organism's individuality, first and foremost because it testifies to its unique fashion of being. The organism's individuality, and in distinction to the identity of inert matter, cannot persist as itself by remaining physically identical with itself. If, for the physicist, "the identity of a material object is . . . the collection of the identities of all its simultaneous member units while they travel together through space-time," then were we to apply a similar principle to the organic, it would spell its immediate and incontrovertible demise.[35] The organic can only maintain itself on the condition that it not remain identical with itself (from the perspective of physical theory). Instead, its only recourse is to participate in a constant exchange of matter and energy with its environment:

> Thus we are faced with the ontological (not only epistemological) fact of an identity totally different from physical identity, yet grounded in transactions among substrates of physical identity. . . . There seems to obtain a rather paradoxical situation. On the one hand the living being *is* a composite matter, and at the same time its reality totally coincides with its contemporary stuff. . . . On the other hand, it is not identity with this or any such simultaneous total, and its reality is not bound to the assemblage making it up now, as this is forever vanishing downstream in the flow of exchange; in this respect it is different from its stuff and not the sum of it. We have thus the case of a substantial entity enjoying a sort of *freedom* with respect to its own substance, an independence from that same matter of which it nonetheless wholly consists. However, though independent of the sameness of this matter, it is dependent on the change of it, on its progressing permanently and sufficiently, and there is no freedom in this. Thus the exercise of the freedom which the living thing enjoys is rather a stern *necessity.* This necessity we call "need," which obtains only where existence is unassured and identity a continual task.[36]

The being of the organism, which is physically manifest in the metabolic exchange from which it is inseparable, is nevertheless not simply reducible to this exchange—at least not if we persist in analyzing it as a mere collection of unique particles. Instead, that which maintains the organism *is* the exchange, for which "necessity for replacement . . . in the organism . . . is as concurrent with the act of existing (because arising from it)."[37] In other words, although we can study metabolic interactions from a strictly external view, if we want to understand the organism in its individuality, then what we must come to terms with is the being of this exchange for the organism. This mode of being, which is no longer simply physical, is nevertheless both physically and ontologically dependent on the physicochemical. In the translation of the physicochemical to the ontological, what emerges is need—that is, the now-felt dependence of the newly inaugurated being on what it is not:

> For the organism the mark of excellence is not immunity to wear, but efficiency of repair—nay, even the very variety and scope of the *needs* which are generated by its existing as such. For its existence *is* performance, whereas for the ship [of Theseus, i.e., an inert object] existence and performance are two different things.[38]

> It is this aspect of "need" which gets us beyond the indifference of the mere form-matter relation. We said before that the "freedom" which the living, i.e. metabolizing, entity enjoys relative to matter is matched by the necessity to exercise that freedom. This necessity is its need, namely, the need for constant self-renewal, and thus need for the matter required in that renewal, and thus need for "world."[39]

"The organism," in sum, for Jonas, *is* the generation of its need, which implies both the relation to a "world" and a form of self-concern. What metabolism names is therefore the very possibility both of need and freedom, achieved through the constant interaction with a material reality (or world) that is, first and foremost, a reality for it:

> For the organism, to exist, in fact, *is* to use and keep using "something else"—therefore to need it. The dependence here in force is the cost incurred by primeval substance in venturing upon the career of organic, that is, self-constituting, identity, in the place of merely inert persistence. Because of the metabolizing mode of this identity, need is of its very essence and it is thus with it from the beginning. But the very *means* evolved to serve this need, such as perception, locomotion, desire, add new kinds of need to the basic, metabolic one—and therewith new kinds of satisfactions. This dialectic too is of the essence of life. The spectrum of needs belonging to an organism is the measure of the degree of its freedom and individuality.[40]

I have been arguing, by way of Jonas, that it is the prosthetic character of basic metabolic activity that on the one hand is responsible for the establishment of organic individuality and that on the other hand destines this individuality to a life of supplements, including both those of mortal and immortal nature. Jonas himself calls this activity, constitutive of life, a performance—not to denigrate it but to give it the ontological priority that is due to what precedes and makes possible empirically verifiable reality. As we see in the passage above, he also links basic metabolic activity to emerging new kinds of being, such as "perception, locomotion, [and] desire." These are, in his words, also elements of the "career of [the] organic," parts of the "spectrum of needs" that evolve as "*means* . . . to serve this need." Although metabolism is thus only the baseline instance of individuality—its minimal necessary and sufficient condition—it nevertheless establishes the possibility for latent, higher-order developments even if it does not anticipate these developments, or the novel possibilities that they in turn inaugurate. In this way, does not metabolism serve Jonas precisely as a form of *différance?*

To conclude, let us finally ask how we might go from (mere mortal) metabolism to indefinite, or even infinite, life:

> The mere phenomenon of *metabolism,* often considered to be no more than the elementary level that underlies, and supplies energy for, the quite different higher functions of life (such as perception, locomotion, and desire), in fact contains in its own primary constitution the groundwork as it were of all those functions, which may or may not evolve on its basis. It presages them by enacting *within itself* the cardinal polarities which those functions will expand and span with their more determinate relationships: the *polarity of being and not-being, of self and world, of freedom and necessity.*
>
> The introduction of the term "self," unavoidable in any description of the most elementary instance of life, indicates the emergence, with life as such, of internal identity—and hence, as one with that emergence, its self-isolation as well from all the rest of reality. Profound singleness and heterogeneity within a universe of homogeneously interrelated existence mark the selfhood of organism. . . . The challenge of "self-hood" qualifies

> everything beyond the boundaries of the organism as foreign and somehow opposite: as "world," within which, by which, and against which it is committed to maintain itself. Without this universal counterpart of "other," there would be no "self."[41]

Metabolism, insofar as it *is*, as the being of organic functioning, always already involves the recursive projection of self, and with self, other. Now, far from wishing to impute an anthropomorphic immortalist fantasy (or self-representation) on the amoeba, the planarian, or the slime mold—each of which, it must be noted, already bears forms of indefinite life, whether it knows it or not—what runs through each of these beings, according to Jonas, is nevertheless some sense of self, even if an unconscious and unminded one. The meaning of metabolism, for Jonas, is just this. As I have already pointed out, this selfhood is, at best, a causally efficacious, artifactual performance that allows for homeostatic and autopoietic acts of self-maintenance to function; at worst, it is a lie that blinds us not only to the essentially sympoietic character shared by all life but also to the plurality of modes of being of organisms, which we risk not only neglecting but also effacing by making recourse to potentially universalizing categories of selfhood. Nor is there any sense in attempting to legislate between these alternatives, which probably are both the case. The point is instead that just as soon as there is minimal articulation of individuality, now understood to be the necessary condition for the autopoietic self-maintenance indicative of life, there is at least the possibility of this difference's being extracted, expanded, extended, and projected beyond every conceivable limit, and as the very expression of the logic of limits. It is this minimal difference that, to all apparent evidence, was present from the very beginning, in the first line drawn by the first inanimate abiotic particles as they achieved, against all conceivable odds, the impossible itself, in a reproducible compartmentalization that would, some 3.8 billion years later, still be an object of fascination for what would come to be known as the life sciences. This line, or trace of difference, can be extended, which is to say replicated, reproduced, cited, supplemented, and iterated. "Extension" does not here name a mode of being opposed to these others but instead a form of interpretation of iteration, supplementation, citation, and so on, as maintaining homogenous (undivided) continuity with their (selfsame) source. Just as soon as there is a being able to be informed by its own, singular difference, just as soon as this singular difference is able to be taken up by a being within what we call the history of life, articulation as trace opens itself to this: its immunitary interpretation as extension. Such a possibility is just as irreducible as its opposite. Nor would it make any sense to attempt to immunize ourselves against this initial immunity, which is a necessary feature of life. What we call mortality and immortality coemerge in this difference that is not one. In both their ontical and ontological interpretations, they articulate formal possibilities of response dictated by the structure of articulation, which is why they manifest at every level of organization and as the phantasm of teleology itself. In Jonas's words, "Concerned with its being, engaged in the business of it, it must *for the sake* of this being let go of it as it is now so as to lay hold of it as it will be. Its continuation is always more than mere preservation. Organic individuality is achieved in the face of otherness, as its own ever challenged goal, and is thus teleological."[42]

PART IV

IMMORTALITY AS TRANSLATION

Two (and a Half) Paths

In Part II, I introduced the term "indefinite life" as a modification of and replacement for the term "biological immortality." The necessity of doing so followed from the fact that today, one imagines that there are at least two viable pathways for realizing human immortalization. According to Nick Bostrom, founder of the Future of Humanity Institute at Oxford University, these two paths consist in the following: "A superintelligence could give us indefinite lifespan, either by stopping and reversing the aging process through the use of nanomedicine, or by offering us the option to upload ourselves."[1] Age reversal and whole brain emulation are today recognized as the two major, techno-scientifically accredited means of radically extending human life. In this way, they speak not only to what is technoscientifically projectible according to the standards provided by the biological, physical, and informational sciences—nothing, according to Bostrom, in principle precludes these developments from being realized in fact—but also to what are conceived of as the necessary and sufficient conditions for continuing our lives. In both cases, therefore, is revealed what makes a life worth living, either through the mode of sheer extension (aging) or that of reduction to a bare minimum (uploading).[2]

Love them or hate them, these two modes are today undeniable in their pull. Even for those unwilling to entertain one or the other as a possibility (be it on philosophical, technical, or ethical grounds), they remain ubiquitous enough to require refutation. For this reason, the question as to whether either is in fact realizable is today of far less consequence than the underlying conceptions of selfhood that subtend each and make either recognizable as theoretical or speculative possibilities in the first place. Even if, in Bostrom's own words, it would take a deity-like superintelligence to give them to humanity, this detail (evidently) in no way detracts from their claim on us, as imagined possibilities. Just as soon as either becomes minimally conceivable, and irrespective of the likelihood of their actual implementation, they become actionable concepts for the sake of which one may, for example, decide to fund artificial intelligence research, or humanity institutes.[3]

In a certain sense, then, age reversal and brain uploading are today undeniable, and their undeniability has specifically to do with their relation to already established philosophical and biological dualisms. On the philosophical side is the legacy of Descartes's substance dualism, which explicitly divides body from mind. On the biological side is the life sciences' twofold manner of studying life, first through neo-Darwinian and computational approaches to populations and genetics, which emphasize the role of information, and second through organism-forward approaches to living systems, which emphasize function.[4] Thus, while age reversal treats the problem of human mortality by attempting to extend the life of the individual as an embodied organism, whole brain emulation does so by forgoing the biological substrate altogether and reducing—or translating—it into what is supposed to be its most essential form: mind, person, or personality (take your pick), each of which is now understood to be the emergent property of an essentially informational—and therefore reproducible—structure.[5] Naturally, the informational structure in this case is not simply that of DNA, but that of some brain, the total organization of which, it is thought, can somehow be discovered, then encoded, before being reproduced.

Given all this, we can at least provisionally distinguish between age reversal and brain uploading at the level of implementation. As prosthetic means of immortalization, they are enacted in divergent fashions, and the nature of this divergence can be comprehended through their respective understandings of what marks the true essence of life. We must be cautious, however, before presuming that such a difference of method and understanding also corresponds to a difference of aim. Just because one credits the first approach, or the second, as being more realistic or more robust in its vision of human life does not entail that the end at which one or the other aims differs in its essence. Insofar as both are appealed to as pathways forward for human immortalization, the principal difference between them comes down to a definition of the human or posthuman (to be immortalized), but not necessarily to a difference in the conceits of that immortalization. In both cases, what one aims at remains an extension of the autonomous, willing, and experiencing self or individual. Where differences begin to emerge, then, is the point at which one's taste for reduction ends.

The figure of indefinite life responds to the necessity of viewing each of these efforts together, as part and parcel of the same—and part and parcel of the same even if, in their approaches to the problem of death, they are anything but. The first and most essential immortalizing prosthesis has never been the technical appendage that comes from the outside and modifies an already constituted individual. On the contrary, and as twentieth- and twenty-first-century visions make all too apparent, the first prosthesis of immortality has always been the self, itself, in its essentially modifiable—and extendable—determination as a present.

Although reversal and brain uploading are by far the two most significant pathways currently proposed for human immortalization—above all by scientifically minded tech elites—we must at least provisionally consider a third: what, more precisely, I call a half path. This third half path is cloning. The semantic ambiguity of the clone, which can refer today both to a biologically continuous extension of an organism and to an informational facsimile (such as a cloned phone) allows the concept to be integrated into either of the paths mentioned above. We might think of the clone, therefore, as a mediate form between the organismal and the informational—a form between the processual continuity of metabolic performance and the qualitative identities of token-type relations. The clone can therefore speak to the informational/organismal divides both internal to biology and between the biological and the technoscientific. Through the various manifestations that it takes—for example, in the fictions of Houellebecq and Ishiguro—the clone reveals the deep continuities that persist between age reversal and brain uploading.

14

AGE REVERSAL

> Setsuko Kashiwara celebrates today her 169th birthday, the most advanced age ever yet attained by a human being. . . . Early this afternoon Setsuko Kashiwara revealed her annual check-up results. At the age of 169, no sign of decline or anomaly. It's a special night, reminding us that we have conquered death.
>
> —Sébastian Mounier, *Ad Vitam*

It goes without saying that the most direct, most intuitive, and most undeniable path to human immortalization lies in age reversal. Age reversal—or senescence stabilization—merely consists in the continuation of a life that, to all appearances, already exists. It is a continuation of an already given life, and one that continues that life in the fashion of its already given existence. It therefore relies only on the belief that *I am* (now), and therefore *I may be* (then). Though it can raise questions of personal identity of the typical ship of Theseus variety—in a hundred, a thousand, or ten thousand years, given the degree of physical, psychological, mnemonic, and dispositional change I am bound to undergo, will I still be myself?—such questions ultimately only reaffirm the idea that *I am* (now), and as a consequence presume, at the very least, the possibility that *I might survive* (in the future). In other words, in no way do interrogations of vertical or horizontal identity, or of Parfitian Relation R (discussed in chapter 15), call into question the basic assumption of the possibility of personal survival, even if they may recast the nature of the survival thereby preserved.[1] The critical thing to recognize, then, is that just as soon as this assumption is set in place, the difference between a successful and unsuccessful survival of the individual is reduced to a sheer technical matter. Its possibility is granted in advance.

Given that there is, to this point, no evidence of a member of *Homo sapiens* living past the age of 130 or so, it seems clear that in order to make biological immortality an accessible state or a technical possibility, certain changes will have to be made to the physical makeup of individual humans. Whether these changes consist in some combination of genetic modification, organ transplantation, nanomedicine, caloric restriction, or sirtuin stimulation is anyone's guess.[2] The point is that some sort of change will necessarily have to be made; some prosthesis will have to come and supplement, transform, or "enhance" the current genetic and epigenetic constitution of *Homo sapiens*. The more extreme the modifications and the greater their number, the more questions of personal identity will inevitably arise. Eventually, although age reversal appears to have an altogether different basis from brain uploading, it nevertheless also begins to raise similar issues. At bottom, however, its allure lies in its holistic approach to life, which suppresses probing questions regarding the nature of, and relationship between, personal identity, consciousness, selfhood, and the body. Age reversal relies on the presumption of some continuity of self. Without this presumption, the fantasy of a meaningful biological immortality loses its appeal and reveals mere metabolic continuity—be it of a part or a whole—the ambivalent implications of which we have already explored.

The problem of self-identity thus haunts age reversal just as it does brain uploading. Although this worry is always latently present for the former, arising sooner or later, at its essence, the true appeal of age reversal is this: it need not raise such questions, allowing them to remain unposed, on the grounds that it merely continues (functionally speaking) a life (in both biological and phenomenological senses of this term) that is (evidently) already in place. Age reversal allows the difficult-to-disentangle relationship between biological and phenomenological life to remain entangled. In this way, although it does still depend on implicit notions of self-identity (just as brain uploading does), unlike brain uploading, the power of the promise of age reversal is that it allows these notions to remain implicit and unspoken.

15

WHOLE BRAIN EMULATION

> If emulation of particular brains is possible and affordable, and if concerns about individual identity can be met, such emulation would enable back-up copies and "digital immortality." "Personal identity emulation": "The emulation is correctly described as a continuation of the original mind; either as numerically the same person, or as a surviving continuer thereof."
>
> —Anders Sandberg and Nick Bostrom, *Whole Brain Emulation: A Roadmap*

> Sometimes VR will place us in other versions of ordinary physical reality. Sometimes it will immerse us in worlds entirely new. People will enter some worlds temporarily for work or for pleasure. Perhaps Apple will have its own workplace world, with special protections so that no one can leak its latest Reality system under development. NASA will set up a world with spaceships in which people can explore the galaxy at faster-than-light speed. Other worlds will be worlds in which people can live indefinitely.
>
> —David Chalmers, *Reality+*

Whole brain emulation offers the dream of a wholly informational, substrate-independent, and thus infinitely reproducible cyborg self. It takes its point of departure in the identification of an essence of the individual—in the mind by way of the brain—from which point it becomes possible to imagine a new substrate on which that individual's program or software can be run:[1]

> By analogy with a software emulator, we can say that a *brain emulator* is software (and possibly dedicated non-brain hardware) that models the states and functional dynamics of a brain at a relatively fine-grained level of detail.
>
> In particular, a *mind emulator* is a brain emulator that is detailed and correct enough to produce the phenomenological effects of a mind.[2]

While age reversal approaches immortalization as a functional problem of organismal physiology—taking as a given that if one preserves the body, then that which matters of self or consciousness will (or at least may) also be preserved—brain emulation approaches immortalization by reducing the individual to one level of its organization: conscious cognitive activity. It imagines not only that this level is the one that truly matters but also that it is in practice isolatable from all the others.[3]

Thus, while both brain emulation (or uploading) and age reversal are presented as solutions to the problem of death, in fact, each of these paths clearly appeals to a different—even if related—notion of life. The appeal of biological immortality today, whether or not it is in fact possible or feasible for us, is our ready acceptance of the embodied, physiological character of our lives. We are but meat machines, as Max Tegmark is fond of putting it, so it is a simple, logical step to hope merely to extend the functioning of what already functions—just a little bit, or a lot, or indefinitely longer. However, for those of us not willing, or not yet sufficiently convinced, that our thumbs or our toes, or even our digestive tracts or our immune systems or our sleep cycles, are essential attributes of who we are—for those, in other words, who are not exactly in denial of physiological reality but who are willing to extract a more essential persona irreducible to these goings-on—brain uploading names an algorithmic reduction of cognitive essentials.

I will not here enter debates concerning the Cartesian nature of this latter position—or, more precisely, concerning the detrimental ramifications of its Cartesianism, and everything that it ignores about both the embodied nature of cognition and the very real possibility that life, consciousness, or both might be substrate dependent.[4] Wherever we may ultimately fall on the problematic reduction of the brain to a machine—or, better, on the problematic reduction of the self to a mind to a brain and then to an (abiotic) machine or program—this point nevertheless stands: there are many today who are all too ready to accept the possibility of these translations as identities. Brain uploading, in strict contrast with age reversal, immediately forces one to pose the question of personal identity. Indeed, before it is even a question of personal identity, it forces a decision regarding what of human life is valuable, and hence the proximity between its transhumanist supporters and neoliberal proponents of eugenics.[5] Brain emulation is always already a matter of cost–benefit analysis, of *reduction,* through which calculation cannot but come to pass. The central conceit of brain uploading is that it reveals that for the sake of which one's life would be worth saving.

Perhaps, then, more so even than the analytic reduction that it forces on one, it is the brazen ruthlessness of the culture of brain uploading that today repulses. While age reversal allows for the deferral of the question of the essence, or value, of human life—indeed, as I have been arguing, that is its primary virtue as a speculative possibility—brain uploading always decides this matter; it decides to decide this matter. The difference between these approaches is thus as much dispositional as it is methodological. There is not only an air of violence but also, perhaps, of vulgarity, if not tawdriness, about the willingness of those who are eager to proffer such a translation of vitality into bits. Yet before it is even a matter of bits per se, already in the willingness to embrace the possibility of translation, for the human, there is an element of this reduction to come. Blatant openness to the possibility of the translation of the human offends what Kant calls the dignity of humankind, and it does so whether or not any of these translations are feasible, or even possible, on technical or practical grounds.

Thus, whether or not brain uploading is possible, and whatever the true relation may turn out to be between generations of humans and their various digital translations, avatars, or replicants, it is above all the faith of our scientists, our tech moguls, and our public intellectuals in the possibility of these translations, between the biological and the informational, that must be gauged. When skeptics decry the reduction of biology to information, they are not wrong, but they miss the essential point. The question is not whether the human—that old, vague, broken category—could ever possibly be perfectly emulated in an inorganic, neural network. (How, for that matter, would we even know? How can I know whether my replica is or truly approximates me? We will come back to this problem shortly.) The question instead is, who is willing to accept this facsimile or reproduction as equivalent, or as more or less equivalent, to the real (i.e., organic embodied) thing? Such an affirmation, however well or poorly justified it may be, is always, at bottom, an act of faith—and an act of faith with very real ramifications, whether or not one is convinced by it.

What calls for analysis, then, is the entire system of accreditation that today allows for the interpretation of this act of faith as one of reason. Even if this system is, to a certain extent, undeniable, it nevertheless remains analyzable. We thus have at least two problems that we must now confront vis-à-vis whole brain emulation: first, the nature of the decision on *who I am,* which I will call the testimony problem, and second, the system that dictates, from the outside, the possibility of the translation of the self, into the other, which I will call the software problem.

Brain uploading approaches questions of life and identity from the point of view of software and hardware. On this basis, its proponents would like to reduce what I call the testimony problem to a software problem. If, in sum, "I" (qua mind or consciousness) am merely the emergent property of an elaborate software program, then reproducing "me" requires only rebooting or reproducing the system. The difficulty is that when it comes to matters of life and identity, testimony cannot be reduced to mere effects of substrate, even if the possibility of testimony (or the decision on *who I am*) is always bound to some material ground. Testimony, in sum, is never merely a matter of an activated program. However, one can always try to reduce the former to the latter. No less critically, one can always opt into the identity between self and reproduction of self, because the nature of testimony is to elude external standards of proof, or disproof. Brain uploading thrives on the possibility of this identity (which is not one) by eliding the question of testimony altogether, appealing instead to so-called empirical standards of objective equivalency. Until, however, one recognizes that the power of this false equivalency ultimately does not depend on the calculations that appear to undergird it, one will be helpless to dismantle its force.

Thus, if (as one appears all too ready to admit today) "I" am merely the persistence of a program (be it genetic, cognitive, or metabolic) that generates a feeling, speaking, reflecting "I," then (it seems to follow) "I" am already a facsimile. It is precisely at this point that the difference between the real and virtual "me" begins to break down, and the reduction of my phenomenal self to its physical substrate seems justified. However, the real issue is that the decision that we base on the appearance of this justification—the decision, for example, as to whether "I" survive in teletransportation—is not susceptible to justification, either for or against. Once we admit this, it changes the nature of the debate on immortality. Interestingly, it does not necessarily foreclose its possibility, but it does disclose its true stakes. So if we—as individuals, collectives, and societies—decide to bequeath our future to colonies of cyberbrains, then, in the words of Marvin Minsky, "[they] will inherit the earth."[6] They will inherit it not because they either are or are not us, but simply because, believing that they bear us beyond our biological lives, we will have given the earth to them.

The problem today is thus no more technical in nature than it was twenty-eight hundred years ago; then, too, it was a matter of autohetero-translation, of the self, in the other. Perhaps the stakes of such translative acts have changed, as they no longer promise merely the survival of the name but now of the autonomous, feeling, willing, desiring self, itself. Whatever the stakes, the necessary passage through externalizing prostheses remains. Just as soon as it is a question of prosthesis, there can be no simple identity between original and translation; no simple identity between the "I" that projects and the "I" destined to bear the weight, or inheritance, of that projection, if only because just as soon as there is prosthesis, the always already prosthetic character of the preprosthetic cannot but rear its ugly head. The difficulty is that it is the originarily prosthetic character of the "I" that spells both the possibility and the impossibility of what we call immortality.

Iterations of Emulation

I have proposed that, on the basis of the originarily prosthetic character of individuality, no form of future identification, of the self, with the other, can be excluded in advance. Be that as it may, given the current state of technoscientific conceptions of self, there are certain fictions of identification that today ring truer, more credible, and thus more valuable or desirable, than others. The two most relevant fictions today are age reversal and brain uploading. Although we have begun to examine the bases on which either age reversal or brain emulation might be affirmed—that is, the reasons why one might come to prefer the one or the other—there remains a great deal of variation within what is known as brain emulation, to which I now turn.

Destructive, gradual, and nondestructive uploading constitute, according to renowned philosopher of mind David Chalmers, the three principal forms that migration of the brain to a nonbiological substrate can take. In exploring these forms, I suspend the question as to the plausibility of substrate independence when it comes either to biological life or to consciousness. A brief discussion as to why I've done so will prove helpful, however, before we proceed.

The question as to the substrate dependence or independence of life and consciousness is important. However, although evidence of substrate dependence could indeed force a revision of the basic premise of brain uploading—which must presume the a priori possibility of the replacement of individual parts constituting the brain, from which simple and easily accepted possibility it then follows analytically, almost like sleight of hand, that the whole must also be replaceable—while approaches to substrate dependence could indeed force a revision of the basic premise of brain uploading, it is nevertheless unlikely that such approaches could actually settle the issues that are put forward in Chalmers's, or Bostrom and Sandberg's, visions of whole brain emulation. The reasons why it is unlikely that they could settle these issues are as follows. First, in proposing not simply simulation but also emulation of the brain, futurists imagine a time when these emulations might be built starting at the level of the individual neuron, atom, quark, and even bit.[7] However ludicrous this may sound, in principle, the driving force of their arguments comes from sheer logical possibility. This sheer logical possibility is then subtended by two further, and difficult to exclude, hypotheses, the first being the future existence of a superintelligence with capacities that far outrun our own, or even anything that we can imagine ourselves as being capable of, and the second being the quantum computational basis of physical reality, or the "it from bit" thesis, which physicists such as David Deutsch, Seth Lloyd, and Paola Zizzi propose as a way of understanding the underlying basis of atomic- and subatomic-level particles. What results from this double thesis—which, again, is difficult to exclude in principle, even if it is no less difficult simply to accept as a given (see Hazlitt and Malthus)—is an emulation that would recreate the specific, physicochemical interactions currently achieved by organic molecules. If the proposal of proponents of brain uploading is to emulate the basic components of organic matter itself, because organic matter (and indeed all matter) is in truth derivable from subatomic, quantum constructors—well, then it is difficult to conceive of a substrate that would not be theoretically reproducible. Indeed, at this point, substrate dependence itself starts to lose its very sense.

Naturally, one can refuse this entire line of argument by refuting the distinctly posthuman premise of a universal computability of matter. That is a possible and perhaps necessary response. However, it is difficult to anticipate the form that such a refutation would take if it is to exclude, in advance, the logical possibility of what Chalmers and Bostrom propose. All they need if they are to continue to promulgate their fictions is the mere nonexclusion of the possibility of such a promise. For this reason, I will entertain the hypothesis of substrate independence for life and for consciousness, if only to see where it leads.

The second reason that knowledge of substrate dependence is unlikely to settle these issues is, even if one accepts none of what I've just said concerning "it from bit," well, at bottom, what is in fact of greater significance than the strong emulation thesis—which it is unlikely we could ever prove we have achieved anyway, given that we cannot at present even prove that you or I am conscious—what is more significant than the argument from possibility versus impossibility is my willingness to endow my digital avatar—whatever its ontological character may in fact be—with preserving and continuing my life. Today one can easily imagine a scenario in which various nation-states juridically authorize forms of self-inheritance, whereby all of one's earthly belongings are bequeathed to a digital AI incarnation. Henceforth I call this the undead billionaire hypothesis. At this point, what matters less than the identity between original and translation is the willingness of both individuals and collectives to recognize, and juridically validate, such identities.

What we need to examine are thus the assumptions that go into this endowment, as well as the assumptions that go into the act of endowing. The whole conceptual, philosophical, and technoscientific apparatus of brain uploading concerns shoring up the essentially performative nature of this act, transforming it from one of faith into one of reason. That is the gambit of everything that follows, and one found alike in the work of Chalmers, Bostrom, and Parfit. It is also why I leave unspecified whether the target of such acts is life or consciousness—not because there is no difference between the two but because the target is by nature a moving one.

As always, the true concerns of life extension efforts are death, the fear of death, and the possibility of technoscientifically eliminating this fear through prostheses that either efface the prosthetic character of life by locating it in a more essential, preprosthetic way of being (i.e., an immutable soul), or efface the difference between differences by reducing all to prosthesis, so as to be able to believe that life itself is not in truth much more than death (Parfit), and that death, therefore, is nothing that we cannot simply integrate into our plan for an ever more extended life. The problem with both approaches, however, is that they neglect the ontological priority of testimony—that is, the always situated nature of experience, which being prosthetic (and thus nonnatural) through and through, is nevertheless not a mere abstraction but rather the very possibility of any decision on self.

In what, according to Chalmers, consists the forms of brain uploading? Destructive uploading, at one end of the spectrum, announces a form of survival in which the original brain is destroyed, while an "accurate simulation of neural behavior and dynamics," captured through "serial sectioning," is loaded onto a computer model.[8] "The result," writes Chalmers, "might be an emulation of the original brain."[9] Nondestructive uploading, in contrast, achieves the same end, but by virtue of a nanotransfer method of brain imaging that does not destroy the original. Finally, gradual uploading also uses nanotransfer, but now in order to replace original biological components—and their corresponding processes—with nanotechnological devices. The key to this last form of uploading thus lies in its progressive replacement of anatomy and function, starting as it does at the level of the single neuron then working its way up through the highest properties of brain function, such as mind and consciousness. Gradual uploading, unlike destructive and nondestructive uploading, exchanges organic elements for inorganic ones, which may either remain in situ or operate remotely through radio transmitters. In Chalmers's own words:

> Here one or more nanotechnology devices (perhaps tiny robots) are inserted into the brain and attach themselves to a single neuron. Each device learns to simulate the behavior of the associated neuron and also learns about its connectivity. Once it simulates the neuron's behavior well enough, it takes the place of the original neuron, perhaps leaving receptors and effectors in place and offloading the relevant processing to a computer via radiotransmitters. It then moves to other neurons and repeats the procedure, until eventually every neuron has been replaced by an emulation, and perhaps all processing has been offloaded to a computer.[10]

Destructive and nondestructive uploading represent the two most radical forms of brain emulation. I will treat them together because they consist in variations on a single theme: duplication. In distinction to these duplicative forms, then, which are essentially qualitative in nature, gradual uploading attempts to maintain the numerical and qualitative identity of the individual. For this reason, it bears a certain proximity to age reversal. Thus, even if proponents of gradual uploading conceptually dismiss the importance of biology, embracing as they do some form of substrate independence, nevertheless, by maintaining the unicity of the subject, they are able to appeal to the same foundation as proponents of age reversal: the temporal continuity of the physiologically unique individual, exemplified by the singular existence of an embodied (even if now abiotic) form.

The irony of what I have just detailed is that—and as Chalmers himself readily admits—just as soon as one accepts the premise of gradual uploading, and even if this form of brain emulation seems infinitely more desirable than the destructive and nondestructive forms, for reasons that I shall articulate below, one ought also accept that of both destructive and nondestructive uploading, even though this may appear to be much more unappealing.[11] The ultimately deconstructible limit between these forms reveals at least two things: first, that numerical and qualitative forms of identity, while helpful heuristics, may fail in the final analysis to be rigorously distinguishable; and second, that despite all their other differences, in each of the above cases, relations of identity must pass through systems of accreditation. Such systems—the establishment of which every technical manifesto on the possibility of brain emulation or biological immortality can be seen implicitly to serve—are what allow for the phantasms of equivalency and identity to emerge in the first place. In the case of destructive and nondestructive brain uploading, the equivalency is one that exceeds both space and time. In the case of age reversal and gradual uploading, it is one that relies on the continuity of both space and time. The paradoxes that emerge in discussions of teletransportation, or other limit cases of survival, are functions of these systems, which are no less the conditions of possibility than they are the conditions of the impossibility of immortality.[12] This is why, as Chalmers also all but admits, these questions can only ever call to belief—that is, instrumental forms of technoscientific knowledge are inadequate to respond to them. However, just as soon as we admit that they come down to belief, the borders separating the technoscientific from the religious or the poetic promises of immortality all but dissolve.

Destructive and Nondestructive Uploading

Both destructive and nondestructive uploading offer qualitative, but not numerical forms of identity: The "I" here is reproduced there. In the case of destructive uploading, the first me is destroyed in the very act that produces the second me. In the case of nondestructive uploading, the first me survives the birth of the second me. Chalmers offers a helpful explanation of the difference between qualitative and numerical forms of identity:

> We can suppose that I have a perfect identical twin whose brain and body are molecule-for-molecule duplicates of mine. The twin will then be a functional isomorph of me and will have the same conscious states as me. This twin is *qualitatively* identical to me: it has exactly the same qualities as me. But it is not *numerically* identical to me: it is not me. If you kill the twin, I will survive. If you kill me (that is, if you destroy *this* system) and preserve the twin, I will die. The survival of the twin might

> be some consolation to me, but from a self-interested point of view this outcome seems much worse than the alternative.[13]

I call both destructive and nondestructive forms of uploading duplicative because once I have been uploaded, in principle, multiple qualitatively identical versions of me could be produced. Thus, while I may not in fact be doubled in the case of destructive uploading, as Pigliucci points out, it is difficult to see how that fact changes anything in principle between these two forms.[14]

The central issue that is posed by both nongradual forms of uploading is thus that of the break in continuity that embodiment traditionally provides. Before there is any question of issues with branching, the simple fact of being reproduced across substrates presents both the risk and the chance of these forms.

Poetic naming, in its consolatory mode, likewise relies on the possibility of retaining an identity across substrates, even if it is neither a fully qualitative nor numerical identity: from person to name, and from name to name (or names). As Montaigne writes in "On Glory," "There is the name and the thing; the name, it's the voice that remarks and signifies the thing; the name, it's neither a part of the thing nor of the substance, it's a foreign piece joined to the thing and outside of it."[15] Apart from the thing, the name nevertheless attaches itself to it, allowing the former to be remarked and signified. The name, in Montaigne's view, is a technology—indeed, a prosthetic supplement—for recalling the thing in its absence, a form, however impoverished, of reproduction. What often gets lost in discussions of the name's preservative function, however, is that it is already performative, thoroughly historical, and essentially idiomatic, even in its minimal form. Shakespeare's sonnets, no less than Homer's epics or Horace's odes, always served double functions. They do not simply immortalize through *the name*—as if the name (as transparent medium or instrument of immortalization) merely awaited their employment—first and foremost, as texts, they strive to establish credit in the felicity of that immortalization as an accredited possibility. Accepting the differences between name and thing, each author articulates a program of equivalence in which it becomes possible to believe that something of the original is carried on. It is this discursive role of the poetry of immortalization that may now prove instructive.

Despite appearances, the difference between this form of immortality, of the name, and that promised by nongradual brain emulation, does not lie in one of kind. In both cases, it is a matter of translation of a type into a token. Both require credence in a continuity that persists in spite of physical and numerical discontinuities, separating original from replacement. Moreover, both depend on the belief in qualitative identities that can only emerge through the reduction of the original to a part or portion of itself. In the case of poetic naming, the relation that holds is between person (or thing) and name. In the case of nongradual brain emulation, the relation is between person and mind, then mind and brain, then brain and its emulation. The discourses of both brain emulation and poetic immortality thus rely on convincing their subjects of the possibility of the persistence of these identities across space and time.

Where the distinction between them is to be found, then, is instead in the categorial difference between the types of objects to be preserved—that is, between the survival of a mere name, as sign, and that of a being with world, or sign user. In other words, even if each ventures the same kind of qualitative identity (multiplication across space; informational homology capable of manifesting in different tokens of the same type; essentially branching structure) there is, nevertheless, (one will say) a world of difference between the survival of something *weltlos* and something *weltbildend*. There is a world of difference, in sum, between something with and without world. To this, one must respond: yes, absolutely. That is precisely the power and force of what today is promised: an immortality not only of dead memory but of the living capacity for memorization; it is an immortality, in short, of the indefinite capacity to create new memory traces and to experience those traces, just as one does at present.

There is indeed a world of difference between the promise of a future with and without world. Critically, this promise is intelligible no less from the perspective of the discipline of biology than it is from that of phenomenology, or philosophy of mind. This is the secret of its power, and that is why the draw of the cybernetic cannot but call into question all known limits. I do not wish to dispute this. Rather, what I wish to contend is only that the survivals that are promised remain no less a matter of translation (i.e., identity in difference) for the proponent of brain uploading than they were for the poet and the poet's subject. In both cases, "I" must credit an other. "I" must credit another "I" with being me. "I" must take on faith that I persist in what, by definition, I am not and I cannot be.[16] This is the very meaning of uploading, which assumes as its condition of possibility the interruption of present with future self. If it did not assume as much, then it would lose its very sense.

In the classic case of the poetic text, I of course at least admit that I will die, even as I hope that part of me lives on. But in the novel case of computational neuroscience, I now take on credit not only that part of me lives on but also that that part in fact lives, breathes, thinks, and speaks for and as me. I now take on faith, in other words, that I will be there, and being there, that I can then affirm the validity of my judgment, now, from across the abyss of my own (newly temporary, and hence reversible) death—from across the abyss of my own death, in sum, as cryptobiosis.

The conceit of this vision is all the more powerful for its vertiginous effect, as it enacts a kind of secret compact between my present and future selves across space and time. Is this not the secret of its allure? The fact, in other words, that it promises now, today, that I not only will have approved of the upload after the fact, just as I do at present, but also that, by way of this certainty of coherence between present and future, there is even a kind of already ongoing conversation underway between my present preemulated and future postemulated selves? We always decide together. We always sign together. That is what brain uploading must mean, if it is to mean anything.[17]

This fantasy of a twice-signed contract is only a further complication of a structure that is in place already for the antique poetic signature. The promise of an impossible reciprocation between present and future is already the effect of a name that, in spite of its inanimateness, is made to sign, for me, in my absence, thereby validating *nachträglich* my present faith in it. If the name did not promise to sign in this way, it would fail to be a name. In the end, however complex these mutually validating structures may become, however convincing they may appear to be, and however codified a shape they may ultimately take, the crucial point is that nothing can confirm or contradict my affirmation of either identity. In both cases, the basic ascription of identity is made across an unbridgeable abyss. In both cases, the performative fiction that establishes the relation is the act of faith, which is therefore neither true nor false but rather an always latent possibility of relationality. What Chalmers and Bostrom would like to convince themselves of, above all else, is that they have somehow eliminated the fiction that lies at the ground of this identity between myself and my avatar, and in this way have managed to eliminate the threat posed by death by reducing it to a temporary lapse.

Nongradual forms of brain uploading thus require that I have faith that the other me is me. But also, and no less significantly, even once I have established the identity between me1 and me2, they require that the fact of this identity—which is not one, and which is always a matter of subscription and hence is not susceptible to standards of proof or disproof—they require that this nonidentity as identity itself be taken to satisfy my belief that once "I" am gone, "I" will live on.

Nongradual brain uploading therefore requires two sleights of hand: the establishment of identity between myself and the other (myself), and the accrediting of this other, on the basis of this identity that is not one, with carrying me on, such that I no longer mind dying. What kinds of calculations are involved here? What is the nature of the credit that underlies either the first or second sleight? To ask whether I live on in my uploaded self is, in form as well as substance, no different than to ask whether I live on in my name, my children, my works, or my deeds. The case of whole brain emulation is only the most extreme radicalization of this classic—and unanswerable—question. Where it appears to diverge from this tradition is in its conceit that what is now preserved is my very ipseity, the presence of which not only can be testified to by the other (as was already the case for the poem, the work, and the child, each of which carried part of me on, but only for the other, since they were needed precisely because I could no longer be there) but also to which it (as I) can itself, across the abyss of bodies and the very negativity of death, testify. This autoaffective testimonial evidence is the ground zero of whole brain emulation's alleged difference. It is both the ultimate dream of self-survival and the basis on which that survival can be hoped for as a strictly secular, empirically verifiable possibility. Yet it cannot but fail to achieve this difference.

To return to my undead billionaire hypothesis, one can well imagine the day when, through the power of newly ratified legislation, I will be able to bequeath not only my fortune but also the very power over my last will and testament to a trustee who still bears my name, but who does so from a computer or avatar interface tasked with continuing me—to make use now of Ishiguro's figure (as I discuss in more detail in chapter 16). This emulation will, at that point, be me. The dream of Chalmers and Bostrom consists in making possible for themselves the belief not only that this future emulated me (me2) will be able to act in my place (which goes without saying) and to do so convincingly (i.e., be able to pass a kind of personal identity Turing test—which does not go without saying, but which seems at least plausible), but also (and most importantly) that in doing so, nothing will be lost. By nothing, we must understand that the demise of my embodied, corporeal form, the end of my ipseity, is reduced to an insignificant blip, a nonevent within the context of "my life." Hence, their dream consists not only in the fact of the external recognition of an identity between me1 and me2, juridically and socially speaking (the first sleight), but also in the possibility that I can know, now, that I will be it, and it will know that it is still me (the second sleight). Anything short of this, and whole brain emulation becomes just another posthumous legacy, ultimately preserved in the other. It becomes just another poetic signature tasked with carrying me on, without me being there.

So what is the condition of this knowledge, which all technical manuals on the possibility of brain uploading serve helplessly to ground? Or, put otherwise, what is the condition for transforming a mere legacy of the word, act, deed, or child, experienced by the other, into a living legacy, experienced by my newly immortal, or indefinite, second first self? What is the condition for elevating my belief in our equivalence (which "I" can never truly know) into sure knowledge of it? Well, I have to actually be there! I have to live and to experience it! I have to not die! If there is death, if "I" must still die, then all we have achieved is a highly sophisticated, labor-intensive, and surely expensive, living but still posthumous legacy. This technical monstrosity that bears my name may bring me greater solace today than the thought of my children or my portrait or my collected works do, but it will not be of an essentially different nature than any of these other prostheses, so far as the promise of survival is concerned. If I must still die, then the alleged identity between myself and this other is no more or less an act of faith than the identity between myself and my name was. Only the other can witness my death and rebirth, which nongradual uploading always involves, by introducing a cut into the continuity of my being. This does not mean that this dream cannot serve to temper my fear of passing. What it means is that that is just what it serves to do. It serves to offer a suture for the wound that Plato and Diotima were already, some 2,400 years ago, ready to admit into the being of mortal life. For this reason, the question is not whether or not uploading works. The question is, who is ready to accept this fiction? Who is ready to live, perhaps to die, and—if the history of the world is any indication—even to kill for it?

A future is imaginable in which, over time, by observing the uploading of others—the similarity in voice, disposition, knowledge, and so on of my friends, with their uploaded avatars—I might slowly convince myself of the nondeath of the original, and then, on the basis of this observation, even grow to believe that I too will not die. This future is imaginable. Moreover, it is easy to see how this new possibility of survival might provide me with some solace, insofar as it will convince me that it will not be necessary to mourn my future death, as so many other postvital promises have served to do. Nevertheless, what it will not and cannot do is eradicate the ineliminable acts of faith on which the whole apparatus must be built. In the final analysis, who could say whether I, or you, would be right or wrong in finding solace in this future? Who could say? The critical point, however, is that the impossibility of precluding such a possibility (of solace, or living on) cannot be confused with the demonstration, from reason, of its empirical validity as a "solution" to the problem of death.

Gradual Uploading

In distinction to nongradual forms of uploading, gradual uploading does not involve a rupture between original and avatar—or rather it involves a series of ruptures, but by way of its spacing out of these ruptures, none of which is supposed to be absolute, it manages to maintain the appearance of the preservation of numerical identity. Through such preservation, gradual uploading seems to escape the fatal flaws involved in the nongradual forms (e.g., the inability of myself to testify to my displacement in the other). Being that I am there the whole time, just like in age reversal, gradual uploading only takes an already present individual, self, or consciousness and modifies it so as to extend indefinitely the possibility of its perdurance. It merely extends the processes whereby I am:

> If a gradual uploading process is executed correctly, each new component will perfectly emulate the component it replaces, and will interact with both biological and nonbiological components around it in just the same way that the previous component did. So the system will behave in exactly the same way that it would have without the uploading. In fact, if we assume that the system cannot see or hear the uploading, then the system need not notice that any uploading has taken place. Assuming that the original system said that it was conscious, so will the partially uploaded system. The same applies throughout a gradual uploading process, until we are left with a purely nonbiological system.[18]

If one accepts the premise of substrate independence, then gradual uploading becomes more or less commensurate with age reversal. I say more or less because, as Chalmers eventually points out, the analytic possibility of the gradual replacement of parts seems to imply, a priori, the nongradual replacement of parts. In other words, if over the course of ten years it would be possible to replace, one by one, all my biotic neurons with abiotic equivalents, then theoretically speaking, why would we not be able to speed up (not only a little, but absolutely) this process? Just as soon as I admit that I can replace a part while maintaining the whole, it stands to reason that I should simply be able to replace the whole. (And does not the natural turnover of our cells already supply—or appear to supply, anyway—a biological model for just such a process?)

> To make the best case for gradual uploading, we can suppose that the system is active throughout, so that there is consciousness through the entire process. Then we can argue: (i) consciousness will be continuous from moment to moment (replacing a single neuron or a small group will not disrupt continuity of consciousness), (ii) if consciousness is continuous from moment to moment it will be continuous throughout the process, (iii) if consciousness is continuous throughout the process, there will be a single stream of consciousness throughout, (iv) if there is a single stream of consciousness through-

> out, then the original person survives throughout. One could perhaps deny one of the premises, but denying any of them is uncomfortable. My own view is that continuity of consciousness (especially when accompanied by other forms of psychological continuity) is an extremely strong basis for asserting continuation of a person . . . as we upload faster and faster, the limit point is instant destructive uploading, where the whole brain is replaced at once. . . . We might formulate this as an argument for the optimistic view of destructive uploading. Here it is to be understood that both the gradual uploading and instant uploading are destructive in that they destroy the original brain. . . . Still, I am confident that the safest form of uploading is gradual uploading, and I am reasonably confident that gradual uploading is a form of survival. So if at some point in the future I am faced with the choice between uploading and continuing in an increasingly slow biological embodiment, then as long as I have the option of gradual uploading, I will be happy to do so.[19]

In this way, and even to his own alarm, Chalmers effaces the gap between gradual and nongradual forms of uploading. Of course, what is missed in this effacement is the importance of numerical identity—or more precisely, what is missed is the importance of giving testimony to numerical identity (what Chalmers does point to with his reference to the "single stream of consciousness"). It is by way of such testimony that one normally staves off fear of death—that is, above all by not raising the question. If personal identity were a simple empirical matter, then the analytic effacement of time (in the merging of gradual and nongradual uploading) would perhaps be acceptable as a premise. However, the identity here at stake is not strictly empirical; it is not a matter simply of biological or computer software, hardware, or even wetware. It is a matter of testimonial faith—a testimonial faith that in fact takes priority over every material identity, being the source from which such identities can be judged as identities in the first place. How else could one even account for the desire to supplement, extend, or enhance the individual, to say nothing of that for staving off death, which is the very condition for this discourse?[20] By saying this, I am not proposing yet another dualism between body and mind. Rather, I am acknowledging the ontological priority of what Jonas and Heidegger call care, which, being grounded in corporeality, is what first allows access to a physical sphere as objectively empirical. The testimonial, or the susceptibility of life to the experience of life, is the site of a fatal entanglement of the biological and phenomenal, and for this reason, it cannot simply be reduced to an effect of the one or the other.

At times Chalmers seems to recognize this, even reflecting on the foundation of belief that subtends all the other sophisticated philosophical operations:

> My own view is that in this case, it is very plausible that the original system survives. Or at least, it is plausible that insofar as we ordinarily survive over a period of many years, we could survive gradual uploading too. At the very least, as in the case of consciousness, it seems that if gradual uploading happens, most people will become convinced that it is a form of survival. Assuming the systems are isomorphic, they will say that everything seems the same and that they are still present. It will be very unnatural for most people to believe that their friends and families are being killed by the process. Perhaps there will be groups of people who believe that the process either suddenly or gradually kills people without them or others noticing, but it is likely that this belief will come to seem faintly ridiculous.[21]

Ultimately, however, if we accept that we cannot simply efface time without also losing the testimonial "I"—or the experience of continuity of self, for the sake of which all this is even proposed, and it seems, in reading Chalmers and the like, that they too frequently forget that the basis of any discourse on life extension is desire, without which none of this makes any sense even to contemplate as a philosophical problem—if we accept, in sum, that we cannot simply merge gradual and nongradual uploading, because doing so entails eliminating a massively important form of testimony, which is preserved only in the former, and if we additionally accept the premise of substrate independence, then what we are forced to conclude is that gradual uploading is just another form of age reversal.[22]

Epilogue on Parfit and Relation R

If brain emulation reveals the calculations that are always already at work in any decision on living on, then it is the virtue of age reversal to suppress this calculus—to suppress it, even if it already takes place and cannot but take place. Such a situation raises the following question: how are we to understand the faith that is already at work prior to any explicit decision on self—that is, the faith of one in oneself as a present "I"? If gradual brain uploading and age reversal are, in the final analysis, to be understood as continuous forms of autoextension, then it must be the case that the calculus at work in the former already takes place in, and divides, the latter.

Importantly, the grounds for being suspicious of brain uploading are in no way simply applicable to age reversal. Even if there were reasons for suspecting that analogous moments of death already take place for the embodied individual, then this would nevertheless not be sufficient to call into question the juridical, experiential, and social realities of lived individuality. For all these reasons, age reversal cannot but appeal. It is, to a certain, significant extent, undeniable within the biotechnological modernity that I have been tracing—and this even if it proves in practice to be strictly impossible.

The purpose of examining biological discourse has been to demonstrate the differential operations through which ipseity may first emerge for the living, above all through processes of metabolism and autopoiesis. Thus, even if none of us would wish to deny indefinite metabolic livelihood as a viable, and acceptable, means of survival (if not simply of immortalization), we must nevertheless also admit that selfhood is itself already differentially constituted. This makes it neither not worth saving nor equally problematic to forms of brain emulation. It does, however, reveal how the already differential character of ipseity opens the possibility of future prostheses of immortalization. It reveals the originary acts of faith that undergird the self and that, at the same time, set into place the possibility of future extensions of self, if not also future extensions of modes of self-identification. We must be cautious here, however. To say that ipseity has a differential character is not to say that it is nothing. It is not to reduce its difference, somehow, to a nondifference, which is precisely what the analytic reduction of the physical universe, and everything contained within the physical universe, including life and mind, to mere bits threatens to do. The prosthetic character of ipseity is what sets it fatally into motion. What it does not do is imply the simple equivalence of all differences to one another at the expense of every evidence, every complexity, and every structure.

Having explored this problem, I simply want to remark on a perhaps surprising similarity in the thinking of a philosopher such as Derek Parfit, for whom the relation between these two forms of autoextension is rather straightforward: "Ordinary survival is *about as bad as,* or little better than, Teletransportation. *Ordinary survival is about as bad as being destroyed and Replicated.*"[23] The reason that Parfit can conclude this is because he rejects the common standard of personal identity on the grounds of his reductionist position. What matters for him—and what should matter for us, he argues—is instead Relation R, by which is meant psychological connectedness and/or psychological continuity: "I have denied that personal identity is what matters. On my view, what fundamentally matters, in our concern about our own future, is the holding of Relation R, with any cause. This would be what matters even when it does not coincide with personal identity."[24] Taking his reductionist position on personhood and applying it no less to cases of spatial (or substrate) difference than those of temporal difference (that is, both to teletransportation and aging), what becomes clear for him is that in both cases, the positive notion of personal identity can only be an effect of differential relations in what we experience as Relation R.

I cannot here engage with the full scope of Parfit's argument. Nor, in discussing it, do I simply wish to endorse his position on these matters, along with all of the problematic assumptions that he makes.[25] Be that as it may, if I find Parfit's work to be of interest, it is for two reasons. First, Parfit, though operating in a fully reductionist mode, nevertheless arrives at a position that at least appears to be structurally similar to the one I have been articulating. By reducing personal identity to Relation R, immortality is rendered at once possible and impossible (my replicant's survival is deemed to be just as bad as that of my being, here and now; in other words, because "I" am differentially nonselfsame, because I am already the product of an identity formation, my survival can neither be assumed nor precluded in advance simply because its physical basis changes.) This conclusion, though correct at a certain level of the investigation, is only possible because Parfit, like

Chalmers, neglects the ontological priority of individuality as a lived phenomenon. Just like Chalmers, Parfit believes that testimony can be reduced to a program running on any substrate, and on this basis, he believes that teletransportation is possible.

The second reason for my interest in Parfit's work concerns what follows, for him, from this first conclusion. After having leveled everything to Relation R (and apparently shown that normal survival is just as bad as being destroyed and replicated), what remains to be done is actually to convince oneself of this. Thus, similar to Chalmers, Parfit must appeal to his personal belief, and like Chalmers, he does so without acknowledging the significance of this act. The possibility of teletransportation, or whole brain emulation, is assured by philosophical knowledge. Nevertheless, for some reason, we must still convince ourselves of this, and of what follows from it: that we no longer need fear death. Thus, Parfit's work, though invested in the rigor of pure philosophical reasoning, also betrays the irreducible stamps of credit and faith that haunt every discourse on life and life extension. Parfit's *Reasons and Persons* is, and can only be read as, a profound act of faith. It even, at times, attains the pitch of prayer:

> **Liberation from the Self**
>
> The truth is very different from what we are inclined to believe. Even if we are not aware of this, most of us are Non-Reductionists. If we considered my imagined cases, we would be strongly inclined to believe that our continued existence is a deep further fact, distinct from physical and psychological continuity, and a fact that must be all-or-nothing. This is not true. Is the truth depressing? Some may find it so. But I find it liberating, and consoling. When I believe that my existence was a such a further fact, I seemed imprisoned in myself. My self seemed like a glass tunnel, through which I was moving faster every year, and at the end of which there was darkness. When I changed my view, the walls of my glass tunnel disappeared. I now live in the open air. There is still a difference between my life and the lives of other people. But the difference is less. Other people are closer. I am less concerned about the rest of my own life, and more concerned about the lives of others.

When I believed the Non-Reductionist View, I also cared more about my inevitable death. After my death, there will be no one living who will be me. I can now redescribe this fact. Though there will later be many experiences, none of these experiences will be connected to my present experiences by chains of such direct connections as those involved in experience-memory, or in the carrying out of an earlier intention. Some of these future experiences may be related to my present experiences in less direct ways. There will later be some memories about my life. And there may later be thoughts that are influenced by mine, or things done as the result of my advice. My death will break the more direct relations between my present experiences and future experiences, but it will not break various other relations. This is all there is to the fact that there will be no one living who will be me. Now that I have seen this, my death seems to me less bad.[26]

16

CLONING AND THE FICTIONS OF IMMORTALITY

> "So you see what's being asked of you, Klara," Mr. Capaldi said. "You're not being required simply to mimic Josie's outward behavior. You're being asked to continue her for Chrissie. And for everyone who loves Josie."
>
> —Kazuo Ishiguro, *Klara and the Sun*

> Upon returning to San José, I immediately went to bed after consuming a massive dose of soporifics. During the following days, I merely roved about, going room to room in the residence; I was immortal, certainly, but for the moment that didn't change much.
>
> —Michel Houellebecq, *The Possibility of an Island*

The Enduring Question of Fiction

Gradual uploading, as well as age reversal, ostensibly preserve what Chalmers calls the "continuity" or "single stream" of consciousness, which for him is "an extremely strong basis for asserting continuation of a person."[1] Without entering once more into the nitty-gritty of body–mind interactions, and suspending the question as to the precise sense in which something like a single stream of consciousness might be said to exist, we can now at least identify in what lies the appeal of this line of thinking: the phantasm of the unicity of consciousness seems to speak first to that for the sake of which, today, many would choose to save/end their lives, and second to a necessary and sufficient condition for just the kind of testimonial evidence that we found missing in nongradual forms of uploading. By recognizing this, my intention is neither to endorse Chalmers's standard nor to credit it with having a priority outside of its specific historical and cultural milieu. (Indeed, it is certainly no coincidence that Chalmers's discourse emerges during the same era that brain death overtakes cardiorespiratory death as the clinical standard operative in most countries of the world.) I simply wish to acknowledge in what lies this vision's particular appeal, even if it is an impossible vision, and even if it is also highly culturally and historically specific. The single stream of consciousness, as standard for personhood, and by way of personhood, for (testimony to) life (extension), is a fiction of the highest order of importance.

But is it not also the case that one only ever has fictions, be they of mortality, immortality, or indefinite life? If I now designate each of these projections as fictions, it is not to denigrate them. Nor is it meant simply to speak to the fact that, to all apparent evidence, no one has ever seen age reversal or brain uploading in the flesh, as it were. The figure of fiction does not here refer to something that is a mere fiction at present but that might, at some point in the future, become nonfiction. The designation "fiction" follows instead from the necessarily performative nature of the phantasms at work in each of these visions, whether or not they are realized (or realizable). It follows from the inherently futural dimension of every promise of immortality, which can only exist as a virtuality, even when it reaches into the heart of the empirically concrete.

The phantasms at work in these visions consist, variously, in what Chalmers calls the unicity of consciousness and in what we diagnosed as the performative character of basic metabolic activity in Jonas, Maturana, Varela, and others. Nor in saying this—in admitting that individuality, consciousness, or mind come about as the differential projections of metabolic, autopoietic, or homeostatic organization—is the point somehow to diminish the importance of life, for example, as a *mere fiction.* Rather, it is to acknowledge that the prosthetic basis of life, in its very reality and as the condition for the production of reality, preconditions the dispersal of these various forms, and it does so always already beyond the limits of the mortal, organic body in the direction of the immortal, indefinite, poetic, and informational avatars, to which it cannot simply be opposed. Instead of rejecting this epithet, we must reclaim fiction in the name of a new understanding of life, so as to rethink, in a more powerful fashion, all its known and unknown limits. Only from the basis of the shared fiction that constitutes life in its minimal articulation can we begin truly to distinguish between the kinds of grounds that subtend visions of afterlife.

Understanding the irreducibility of fiction (or credit) within the relations that constitute "me" is the key to understanding why prostheses of self, from the name up through the uploaded brain, remain not only virtual possibilities but also to a certain extent inevitable expressions of an inexorable profusion of autoextension, the origin of which lies in nothing less than the impossibility of the self's own undivided identity. One must in each case credit an identity between parts that, by their very nature, cannot be equivalent. This is what Plato already saw: nonidentity is the law. But the failure of the principle of identity, which regulates discussions of immortality from Plato on, does not simply spell the end of immortality. It is at one and the same time immortality's condition of possibility and impossibility. It announces the birth of immortalization as prosthesis, whose laws are no longer regulated by straightforward logical distinctions. Why is that?

On the one hand, non-self-identity dictates that any preservationist effort modeled on presence is doomed to fail. How can I extend what is itself not one? How can I preserve what was never there, in and as a present? Every attempt to make substantial an immortal soul testifies to the inveterate failure of such efforts to locate an object to be saved. As soon as one admits this absence, the classical concept of immortality as infinite, immutable, and indivisible no longer has a leg to stand on. It is doomed in advance.

On the other hand, non-self-identity dictates that whatever I call "I," being already differentially prosthetic in nature, can in fact be so extended, but only as a differential economization of that difference. This does not mean that "I" can live forever. It means that "I" has always already engaged in calculations on itself. It means that "I" is always already a matter of differential translation, and therefore that "I" can always opt to sell itself for this or that fantasy. The problem with critics of transhumanism is that they believe that it actually matters whether the fictions put forward by Parfit, Bostrom, Sandberg, Itskov, Kurzweil, and Moravec go too far and assume too much in taking their digital clones as extensions of themselves. That is, critics too often take for granted that the success of transhumanist fictions depends on the validity of their claims. What philosophical or technical demonstration could possibly reveal the veracity or falsity of such acts of identification? The essentially performative structure of identification is not susceptible to apodictic or correlationist claims of right and wrong. Identification is in each case a profession of faith, and to fight such a profession, one must be willing to see it for what it is.

Accreditation, or identification, is always a matter of faith. This is also why no one can arbitrate between your vision of the afterlife and mine; why no one can dictate whether we ought to prefer to live on through our biological offspring, the imprints of a book, the breath of a clone, or the program of a cyberbrain. On each occasion, accreditation is a matter of faith, even if the decision is always weighed through reasons, and even if we cannot but fabricate reasons to defend (or perhaps obscure) the decision that decides nothing. This is what, as I discuss in this chapter, the novels of Houellebecq and Ishiguro reveal. Living on is an irreducible act of faith, which is to say that at the limit, it is nothing that can be validated or invalidated, either in advance or after the fact.

This latter point brings us, at last, to cloning. Earlier, in Part III, we explored the biological basis of immortality via the binary fission of unicellular organisms. These clones maintained both qualitative and numerical identity with their predecessors, but in doing so, because of the resultant branching, they in fact did more to reveal the limits of our conceptual apparatus than they did to bolster a notion of survival that most would opt for. Though the clones that we now take up remain, in some sense, the descendants of those former ones, they also expand the concept in radical fashion. By establishing their genetic relations to their predecessors through amalgamations of personal and physical features, these clones abandon the contiguity of metonymy for the rupture of metaphor. The interest of the clone lies, precisely, in its ability to claim strong forms of continuity despite this interruption.

In distinction, then, to both age reversal and brain uploading, cloning always presents a fallen form of immortality. The clone is, to adopt Benjamin's parlance, the ruin of those other modes, betraying the decay of their auras and the already fragmented, allegorical nature of their otherwise brilliant, symbolic forms. If I take up the figure of the clone now, then, it is because it helps illustrate the calculations already at work for age reversal and brain uploading, as well as the residual power of their promises, even in the wake of their failures to meet the standards they set for themselves.

Cloning

What is the clone? Well, in brief, the clone *is not*—or, to put it otherwise, there is no predetermined concept of the clone. What the clone represents, today, is rather the possibility of the indefinite extension of the logic of identity in and through media both known and unknown. The clone is the ultimate expression of the irreducibility of faith within relations of identification; it is also the clearest reminder of the failure of such acts to cover over the difference that makes them necessary in the first place.

It is true that according to its etymology, the clone starts out as the branch broken off the tree. It is the first fracture, and as such, it is the first figure for the threat of loss and its correlate hope of recuperation. In its microbiological configuration, it next becomes the basis for affirming the continuity of life now understood as preservation of metabolic and genetic identities. Yet even as the figure for the continuity of the biological, the clone will never have been fully determined by this field. The clone can determine the biological only on the condition of exceeding it, and this excess is borne out by its eventual displacement into the abiotic, the technological, and the commercial. Today one may clone a computer or a phone just as easily as a plant or a pet, even though the technical procedures for each bear no resemblance to each other. Certainly a clone is always a double; but the ground of relation linking clone with cloned is essentially plastic. It appeals at one moment to numerical identity through contiguity, at another to qualitative relations between types and tokens, and at yet others to everything in between.[2]

It is this plastic nature of the clone that lends it so easily to dramas of suspense, always bordering on horror. The reason for this is that the clone has no essence. It posits an identity, but an identity without stable ground. This identity can always take on a new ground for its basis, because what matters in the end is never the nature of the relation between clone and original, but only the former's ability to pass for the latter. At the limit, for the clone to be a clone, the only condition it must fulfill is the semblance of replacing the original, on any grounds.

Thus, Duncan Jones's *Moon,* Christopher Nolan's *The Prestige,* and Kjartansson and Kormákur's *Katla,* by variously confronting the horrors of replaceability, each deals in its own way with this same clonal question. The other me who is me is not me. No matter the basis of identity—be it biological, informational, psychological, or other, and be it grown from my cells or downloaded from the cloud—the testimonial gap in each case remains unbridgeable. Yet—and hence the horror—it is all too easily bridged for the other.

Can this gap not be bridged also for me? Can we not learn to suture this wound? Can I not learn to believe in my rightful immortality, in the other (me), as Plato, Parfit, Bostrom, and Chalmers would each teach? To the precise extent that the fallen figure of the clone ceases to claim a strict identity with the original it replaces, it reveals the fictional bases of the relations subtending fantasies of brain uploading and age reversal. It reveals the line that conjoins and divides the two forms, and that cannot but continue to be modified and amended as the dialectic between technoscientific discovery and autoidentification proliferates novel forms of selfhood and personhood.

Two examples demonstrate this. The first is Kazuo Ishiguro's *Klara and the Sun.* Ishiguro's novel explores what it might mean to "continue" someone—that is, to adopt another's life at the point of their death. Set in a world in which AFs, or artificial friends, are manufactured to supply companionship for isolated (natural born) youths, the novel takes up the family of a young girl, Josie, and her mother, Chrissie. Josie, for not entirely innocent reasons, is threatened by illness.[3] One day, she and her mother decide to purchase an AF named Klara. Though she is artificial—she is solar powered and has a speech pattern that distinguishes her from her organic peers—Klara is quite humanlike and becomes an integral, accepted member of the family. Klara is also sensitive, with a preternatural talent for perceiving and emulating traits such as gait and disposition.

At a decisive moment, Josie is deemed too sick to go on a prized excursion, so Chrissie decides to take Klara with her instead. During their time together, Chrissie makes it clear that the scope of Klara's role may in the end exceed mere companionship:

> "I think it was right we didn't bring Josie today," she said. "She wasn't well. But now we're sitting here like this, I do miss her."

> She looked around, turning to look at the waterfall. Then she turned back and her gaze went past me, to the passers-by, the dogs and AFs. "Okay, Klara. Since Josie isn't here, I want *you* to be Josie. Just for a little while. Since we're up here."[4]

In asking Klara "to be Josie," Chrissie is initially asking for an imitation. As Klara adopts her role, Chrissie, the mother, addresses Klara as Josie, the daughter, and goes on to discuss with her the reasons why she couldn't have taken her on this trip. If this is an early experiment in Klara's capacity to convincingly imitate Josie—although in point of fact this had been the first task given to Klara, still at the store before she was purchased—then, we eventually learn, it is in fact also part of a broader plan, and one that Josie herself is eventually made complicit in: for Klara to continue Josie, just in case Josie herself cannot go on. In this case, what begins as a mere imitation is intended to become something else, something both stranger and stronger. Josie has been sitting for a "portrait," and Mr. Capaldi, the "artist," intends to use it, along with Klara, to create a convincing continuation of Josie that will transcend mere mimicry:

> "What you have to understand is this. This new Josie won't be an imitation. She *really will be Josie.* A *continuation* of Josie."
> "You want me to believe that? Do *you* believe that?"
> "I do believe it. With everything I'm worth, I believe it." . . .
> "But will I believe in it? When the day comes. Will I really?"[5]

> "Klara, we're not asking you to train the new Josie. We're asking you to *become* her. That Josie you saw up there, as you noticed, is empty. If the day comes—I hope it doesn't, but if it does—we want you to inhabit that Josie up there with everything you've learned. . . . So you see what's being asked of you, Klara," Mr. Capaldi said. "You're not being required simply to mimic Josie's outward behavior. You're being asked to continue her for Chrissie. And for everyone who loves Josie. . . . The second Josie won't be a copy. She'll be exactly the same and you'll have every right to love her just as you love Josie now. It's not faith you need. Only rationality. I had to do it, it was tough but now it works for me just fine. And it will for you."[6]

Klara is meant to become Josie. Though the word is not used, what she is meant to become is her clone—a clone that, being so close to the original, is able to efface the gap between the two. That is what is supposed to happen; it is what is supposed to happen according to an intention that is publicly shared in the novel and that is meant to operate not through deceit but with full conscious awareness. According to Capaldi, the fidelity of Klara's continuation of Josie depends only on the convincing nature of her performance, which is to be so convincing as to turn imitation, or simulation, into full-blown identity.

Within the fiction that is *Klara and the Sun,* what is supposed to matter, as concerns Klara's continuation of Josie, is only Klara's capacity to know and, by way of knowing, to convincingly simulate Josie. Such a performance depends not at all on the physical substrate that supports the AF. The biotic/abiotic nature of Klara is, for all intents and purposes, concealed. There is no link between her and Josie through either DNA or the fallacious analogy of brain emulation. The authenticity of her continuation, of her taking on of Josie's life, is simply to be judged by the other, who will either be convinced by it or not, and who as a result will either come to believe in it or not. Nor is any lie being perpetrated here. There is no conceit that the relationship between Josie and Klara will be seamless for Josie. There is only the compromise position, evidently subtended by rationality, that where there is belief in identity, there is identity.

What is of interest in *Klara and the Sun* is not whether such a vision is convincing for us; of interest is the nature of what is being perpetrated and the fact that it ought to be convincing for them. Ishiguro's novel presents a scenario in which a collective agrees (or tries to agree—or agrees to try to agree, anyway) to the possibility of continuation. In this way, identity, the novel proposes, comes down to a question of belief. Like Kleist's "Foundling"—but perhaps both more and less cynically—Ishiguro's fiction contends that identity itself is a question of credible fictions.

In this way, *Klara and the Sun* explores the plastic nature of the clone and the role of belief in underwriting personal identity. However, if it goes very far in its consideration of the collective as a willing agent in the effacement of difference between original and replacement, it nevertheless leaves mostly open the question of the individual's possible participation in its replacing.[7]

The extreme form of such a compact, in which it is not only the family, the group, or society at large but also the subject itself who agrees with and even comes to believe in the felicity of such a continuation, is found in Michel Houellebecq's *The Possibility of an Island.* As with Chalmers's nongradual brain uploading, Houellebecq's proposed form of survival is not seamless. In *The Possibility of an Island,* immortality is a condition reached only through the promise of a future replacement, after death. Immortality is therefore to be found across events of rupture, when the other takes up one's life after one's death. It is nothing, therefore, that one can testify to, but it is nevertheless something that one can opt into and embrace. It is a possibility of life that may be subscribed to and affirmed, and that one can subscribe to and affirm in the mode of belief. If the primary concern of *Klara and the Sun* remains the possibility of convincing the other of one's survival, then that of *The Possibility of an Island* is now the possibility of convincing oneself. As Houellebecq writes, "In all likelihood the idea was too new. To be honest, I forgot myself most of the time that I had become immortal. It was necessary to make an effort in order to remember."[8]

The Possibility of an Island is a novel told through the multiple voices of what we are led to believe is the same man. On the one hand, there is the commentary of Daniel1, our contemporary and the novel's primary protagonist. On the other hand, there are the commentaries of Daniel24 and Daniel25, the neohuman descendants of Daniel1, who later come to replace and reincarnate him. At the most basic level, *The Possibility of an Island* tells the story of the birth of the Elohim cult from the perspective of Daniel1. It tells the story of the cult's transition from an essentially fringe faction promising an immortality that it can in no way furnish to a global power with sufficient financial backing and technical know-how to spawn genetic clones of its adherents. At a more complex level, *The Possibility of an Island* tells the story of the participation of Daniel1 in this cult, and of his transition from nonbeliever to believer. This transition, which runs contemporary to the cult's financial and technical progress, is the text's true subject, which it explores through the fragmented voices of Daniel's past and future selves.

The immortality that is promised in *The Possibility of an Island* is not one of gradual or absolute nature. In each case, it entails death and rebirth. In Daniel's world, after "I" die, a genetically similar clone will be spawned, and that clone will be tasked with internalizing the commentaries written by each of its prior generations as well as with penning its own. In theory, this testamentary act of writing is intended for the clones that will follow, but in practice, it also serves to inscribe the individual within the communal whole. Each clonal self thus signs the same pact that is signed by all the others, and in so doing, each admits and acquiesces to the fiction of their mutual identity. As I shall observe, this is the process through which a fictional fiction of identity may become a nonfictional fiction of identity.

It is the specific combination of genetic and testimonial continuities between the generations that is supposed to assure the preservation of identity in Houellebecq's fictional world, and as a result to warrant the use of the figure of immortality for the individual. "Testimonial" in this sense does not refer merely to the experience of the sheer continuity of presence in consciousness, like I used it earlier. Instead, it refers to the decision, within the experience of the sheer continuity of presence in consciousness, by each individual, to appropriate the alterity of the other, which it does not experience, as part of the same. The determination of my immortality—or my survival—is consequently nothing that I can know immediately or absolutely. It is instead a knowledge that is only available through complex acts of ratiocination—as, it should be noted, it has always been—and this is surely why Daniel1 tends to forget it.

My immortality—or, to be precise, the promise of my survival in my clone, or my predecessor's survival in me—can therefore only affect me as a form of cultured knowledge, produced through complex discursive operations. In Daniel1's case, these operations consist in the formation of a new cult that marries cutting-edge science with age-old propaganda. Yet while Houellebecq's is an essentially cynical vision, more likely to dissuade than to persuade anyone from embracing such a form of existence, it is cynical in an importantly complex way: *The Possibility of an Island* is about the self-deception inevitable for every dream of life extension, yes, but it is also about the inevitability of such forms of self-deception. What we read when reading the novel is the process through which Daniel1 gradually sells himself the promise of an immortality that is not one—or, at any rate, the promise of an immortality that is one only because he buys into it as one. Self-deception is therefore the currency of this transaction. Yet—and herein lies the supplementary interest of *The Possibility of an Island*—because belief is a fundamental building block of every dream of immortality, which always comes down to questions of credence, to something like the minimal projection "I will be there," be it in a future I cannot know or in the wake of my death, what commences as mere deception does not conclude as such. As external observers, we may of course be primed by Houellebecq to dismiss Daniel1 and his immortalist self-conceit on any number of grounds. Yet such a judgment—on our part—can do nothing either to affect the felicity of his performative declaration—that he is immortal—or the actions that he takes on its basis. Because immortality has never had a stable referent, one may reinvent it on virtually any grounds. Unmasking Daniel's performative for what it is thus cannot serve to invalidate it as a credited possibility. This is, perhaps, the novel's final lesson:

> Fox's [Daniel's dog] genetic code had been preserved, he reminded me, and we had become immortal; we, but also, if we wished it, our pets.
>
> He seemed to believe it; he seemed absolutely to believe it, and I felt suddenly paralyzed with joy. With incredulity, also: I had grown up, I had aged, in the idea of death, and in the certainty of its empire. It's in a foreign state of spirit, as if I had been on the point of waking up into a magical world, that I awaited daybreak. Day broke, colorless, over the ocean; the clouds had disappeared, a miniscule corner of blue sky appeared on the horizon.[9]

Daniel1 becomes a believer in his own immortality. He comes to believe that he will be—indeed, that he already is—immortal, at the very moment when the technical hurdles to human cloning are overcome. Perhaps he does—which is neither to say that he will live forever (he does not) nor to contend that his particular vision is any more or less valid than any other vision, but only to admit that we are here touching on the mystical foundations of a claim that is, to a significant extent, self-grounding. We may opt to reject this claim. We may opt to ratify laws for or against its possibility. We may choose to mobilize arguments and evidence that reveal its implicit conceits to be either true or false; indeed, much of this book has done just that. Yet—and this is the critical point—because immortality has never had a single incontrovertible sense, because the lines of descent that constitute self and other are forever in flux, the possibility of such a claim, however idiomatic, isolated, and insular, remains. Its many variations cannot be anticipated or excluded in advance.

Within the context of the novel, Daniel1's belief is confirmed by Daniel24 and Daniel25, each of whom adopts his view, and in so doing adopts his life as their own:

> When I consider the night sky it occurs to me to think of the Elohimites, of this strange belief that must, finally, through circuitous paths, unleash the Great Transformation. Daniel1 lives again in me, his body knows a new incarnation here, his thoughts are my own, his memories my own; his existence is genuinely prolonged in me, much more than any man has ever dreamed of being prolonged through his descendants. My own life, however, I think about this often, is far from being one that he would have liked to live.[10]

What to make of this fiction? On the one hand, just as in *Klara and the Sun,* what is here offered is a form of continuation, and one that does not promise the continuity of the single stream of consciousness. We would therefore be well justified in asking whether it should still be called immortality. But on the other hand, is not what the fiction of immortality itself serves to perform, on each occasion that it is performed, precisely the single stream—that is, the conceit and legibility of something like a *single* stream, even, and especially, there where there is none? Given that what is on offer here is clearly not a continuous presence of ipseity, so long as one chooses to believe in the existence of something like the single stream, as being truly single, such instances will always remain secondary and derivative. However, just as soon as one admits that, what the single stream of consciousness, like the mutual identification of Daniel1 with Daniel24, and like the so-called sheer contiguity of contact of life itself, dating back 3.8 billion years, each primordially serves to confirm is just the possibility of something single, continuous, and present to itself—*extension*—then the gravity of Houellebecq's vision becomes much more difficult to dismiss.

Where Chalmers, Parfit, and Bostrom each falls short, therefore, is in failing to perceive the fictional limits of their own discourses. Convinced of the validity of the forms of survival they put forward—or, perhaps it would be better to say, invested in demonstrating the validity of the forms of survival that they themselves wish they could fully affirm—they fail to see how they too have become adherents. They too have sold themselves a story of self-continuity that is no more grounded in empirical reality than is Chrissie's or Daniel's. Such a realization, however, does not function simply to spoil the immortality in each case dreamt of. Nor does it necessarily serve to eliminate its allure as such a dream. Instead, it merely situates it within its field of production. Yet whereas the novels of Ishiguro and Houellebecq are still able to reflect on the fictional statuses of their texts, and in this way reveal the performative force of their claims, it is precisely the function of the texts produced by Bostrom, Chalmers, and Parfit to efface the fictions that underlie them. Unfortunately, by effacing these fictions—or, more accurately, by attempting to efface what can only ever be covered over—these texts also blind themselves to the most interesting thing about them: the performative character of their fantastic fabrications.

CONCLUSION

Am I, Then, Immortal?

When does one become immortal? Is it as simple a matter as opposing before with after? As we have seen, the "when" of immortality is never straightforward. Though, to all apparent evidence, we do not yet possess the technical acumen to create an immortal *Homo sapiens,* as Harari was ready to admit, we may yet be walking amid future immortals today. One may therefore already be immortal without knowing it yet. Immortality, in its modern, biotechnological configuration and in its interpretation as indefinite life, is marked through and through with uncertainty, and this uncertainty touches us all.

The question of immortality's uncertainty has, to a certain extent, always been posed. The entire tradition leading from Plato to Descartes confronts it, offering various proofs for immortality's existence, and doing so because it is nothing that can be known. Of course, this sort of nonknowledge goes both ways: the immortality of the soul, or God, is not something that can be shown to be or not be. Uncertainty is therefore an irreducible element of immortality's religious determination, but one that marks it not as temporally indeterminate, but epistemologically so. Thus, whether or not we know about it, and whatever we may or may not accomplish, here and now, for the likes of Plato and Descartes, immortality's status as a reality or a fiction is always already determined. Nothing in this world can change that.

The conditions marking the uncertainty of immortality in its current guise, as indefinite life, have, for this reason, certainly undergone transformations. If Harari, Bostrom, Sandberg, de Grey, Chalmers, and Itskov are right, then, far from already determined, everything in fact still hangs in the balance when it comes to indefinite life. One either will or will not have been immortal, and the force of this either/or now rests on the realization of positive, technical means. Immortality is, for this reason, today largely an ontic matter; it is one to be realized here, in the flesh, in this world. It ought, therefore, to be susceptible to questioning, and, not only that, but also to verification—hence the appeal for many of indefinite life vis-à-vis immortality's earlier, theological determinations.

Immortality is now supposed to be an ontic matter. Nevertheless, as such an ontic matter, it retains ontological implications. It holds consequences for the way that I am and the way that I relate to my being. In other words, if it is the case that I either am or am not immortal, today; if we allow that this is a testable hypothesis, then it nevertheless remains to be seen how this fact of immortality might affect the being of my being or the way that I am in the world. Boredom, nihilism, and meaninglessness, say some; activity, excitement, and total self-realization, say others. The question of immortality's technical feasibility must therefore eventually cash out in the lived experiences of those who either do or do not partake of it. Immortality, in sum, should be *felt.*

I conclude by exploring the question of the being of the new immortal. To be precise, I would like to revisit an approach to this question, and one that largely sidesteps the mostly bankrupt debates concerning the relative advantages or disadvantages of indefinite life, considered as a modification of finitude. I do so by way of a work of fiction by the author of *Frankenstein, or The Modern Prometheus.*

Mary Shelley published the short story "The Mortal Immortal" in 1833, some fifteen years after the incomparable *Frankenstein.* In its opening lines, its narrator announces, without apparent irony or prevarication, that he has just completed his "three hundred and twenty-third year." He goes on, over the course of this short but incredible text, to tell us about the life he has lived, how he came to survive this long, and his hopes for the future. He speaks of love, the errors of youth, and sorrow, thereby speaking of everything that we too might have thought to relate, as though he were indeed still one of us, which is to say, inclined to relate, as well as to write, speculate, regret, reminisce, and project. By the end, his story finished yet with no clear closure provided, we are left to wonder at the marvel of this man, who may very well be just that—a man—or perhaps something else, and so much more. In this way, Shelley's fiction offers us a view onto immortality, but from one who very much remains mortal. Our question, in reading "The Mortal Immortal," concerns what it is like to be in such a position. It concerns what it is like to be between these two states—perhaps destined to live forever, but perhaps not:

> July 16, 1833.—This is a memorable anniversary for me; on it I complete my three hundred and twenty-third year!
>
> The Wandering Jew?—certainly not. More than eighteen centuries have passed over his head. In comparison with him, I am a very young Immortal.
>
> Am I, then, immortal? This is a question which I have asked myself, by day and night, for now three hundred and three years, and yet cannot answer it. I detected a gray hair amidst my brown locks this very day—that surely signifies decay. Yet it may have remained concealed there for three hundred years—for some persons have become entirely white-headed before twenty years of age.
>
> I will tell my story, and my reader shall judge for me.[1]

In this way begins Shelley's text. Her narrator—who, we learn in the ensuing pages, drank an alchemist's elixir, after which, to all appearances, he stopped senescing at the age of twenty—here asks us to judge, for him, as to his immortality. Having once been mortal, the imposition of immortality is something he can well speculate on, but is not, evidently, something that he can know. Being the possible subject of an experience of immortality, here, in this world, thus offers no privileged insight into the nature of that state, or even really into whether one is or is not in that state. "Am I, then, immortal?" he asks, the adverbial "then" marking out the role that observation, reason, and judgment are to play in the formation of any possible response.

The problem of immortality offered in Shelley's "Mortal Immortal" is an insoluble one, and one, moreover, that can be addressed just as well to the other as to oneself. Its speaker asks whether he is immortal. He tells us his story and asks us to judge for him. His question is how he—or anyone, for that matter—can know if he is immortal. It is how he—or anyone—can know if he shall die. Supposing, for a moment, that he is immortal—whatever that might mean—he nevertheless remains haunted by the question of his mortality, as all of us so-called mere mortals are. Supposing that he is mortal, and thus not knowing under what conditions or in what time frame death might surprise him, he is still left questioning his possible immortality. He is, therefore, a mortal immortal to the extent that this aporia haunts—and even constitutes—his being. However, I would propose, if this is the aporia that constitutes his mortal immortality, in the ontological uncertainty of his being, and of the time of his demise, then we too must be mortal immortals, and in just the same way.

We too live in the uncertainty of the date of our demise; we too live with the certainty of the impending nature of the impossible possibility of death, which latter impossible possibility haunts us not in the mode of a when but of an if. We too know not when we shall die, or if we shall die. Such nonknowledge is, perhaps, the novelty of our biotechnological modernity and of the technoscientific configuration of modern prostheses of immortalization. The consequence of these forces is to ontically extend the separation between what Heidegger calls being-toward-death and Derrida the condemnation to die. This separation is therefore growing wider, and its growth is a consequence both of radical transhumanist fictions and of basic research in the life sciences. What it exposes is that the impending impossible possibility of death, which is a general, structural feature of life and world, no longer need be conflated with the predestination to demise as a factical, present event.

Make no mistake: as actuaries are more than happy to tell us, it is a near statistical certainty that we, all of us, will one day cease to be. My discourse is not, in this respect, an inherently optimistic one. It neither predicts nor necessarily supports radical extensions of our lives. Rather, it is, above all, diagnostic, and what requires diagnosing is, first and foremost, the aporia of mortal immortality (as an ontological state), which today finds its most extreme, external reflection in the ongoing and ever-present deconstruction of death at work everywhere—not only in the biomedical, informational, and physical sciences, but also in the literary, televisual, juridical, and commercial spheres.

The separation between being-toward-death and the condemnation to die (what I have called indefinite life and am now calling mortal immortality, and crediting Mary Shelley with having marked it out for us) may, to a certain extent, have always been the case. It may have always been a virtual possibility, and one that we sense every time that we fail to believe in the reality of death, for us.[2] What has changed, therefore, and perhaps irreversibly, is the visibility of this separation, which we now suffer as a coming historical event that we may choose to either embrace or reject. We suffer this possibility of our nondemise no longer as a mere logical possibility but as the threat of a coming reality, which is to say of a coming, newly accessible state.

Am I, then, immortal? Naturally, this is another way of asking, can I die? Can I can still be sure of the certainty of my death, as a present event that comes and that I, perhaps, welcome? The lifting of the "when" of ontic demise is, in this respect, also a kind of loss, as the continued and unabating propagation of HeLa cells constantly reminds us, and as does the revelation of cryptobiosis in the biomedical acceptance of forms of suspended animation. In both cases, what is spelled out is the end of a certain fantasy of sovereign control, and one that finds its correlate today in the threat of the rebirth of my cyborg, clonal, or digital avatars, years, or perhaps eons, after my demise. This, too, is a form of loss.

> Am I immortal? I return to my first question. In the first place, is it not more probable that the beverage of the alchymist was fraught rather with longevity than eternal life? Such is my hope. And then be it remembered, that I only drank *half* of the potion prepared by him. Was not the whole necessary to complete the charm? To have drained half the Elixir of Immortality is but to be half immortal—my For-ever is thus truncated and null.
>
> But again, who shall number the years of the half of eternity? I often try to imagine by what rule the infinite may be divided. Sometimes I fancy age advancing upon me. One gray hair I have found. Fool! Do I lament? Yes, the fear of age and death often creeps coldly into my heart; and the more I live, the more I dread death, even while I abhor life. Such an enigma is man—born to perish—when he wars, as I do, against the established laws of his nature.[3]

Is it the case that I know that I shall die? This question can now be heard with multiple intonations. Not only: do I know that I shall die, do I know that my life shall come to an end, do I know, in sum, that I am condemned to die? But also: can I be sure that the limits of my "I" begin and end with those of my corporeal being? Can I be sure that death is something that happens all at once, irreversibly, and without remainder? Might I not instead live on, and even do so in spite of myself? Mortal immortality names this possibility, no longer as the metaphysical phantasm of immortality (i.e., as elimination of contingency and chance) but now as the possibility that I might live, or live on, and do so without foreseeable end. I don't know that I'm not immortal. I don't know that my life won't be extended, radically, and this non-knowledge shall never be surpassed. Death, let us say, is no longer a given. Nor has it ever been.

NOTES

Foreword

1. Maurice Blanchot, *The Instant of My Death,* and Jacques Derrida, *Demeure: Fiction and Testimony,* trans. Elizabeth Rottenberg (Stanford, Calif.: Stanford University Press, 2000), 5.
2. Jacques Derrida, *The Death Penalty,* vol. 1, trans. Peggy Kamuf (Chicago: University of Chicago Press, 2014), 1:219–20.
3. In 2015, Glossip was the petitioner in an important Supreme Court case (*Glossip v. Gross,* 576 U.S.) relating to the Eighth Amendment's prohibition against cruel and unusual punishment. It involved the constitutionality of Oklahoma's lethal injection protocol: 500 mg of midazolam, followed by a paralytic agent, followed by potassium chloride. In short, it sought to address whether the midazolam prevented the pain inflicted by the second and third drugs. The Court found that it did, by a 5–4 majority. In dissenting, Justice Sotomayor famously suggested that the condemned person might feel "the chemical equivalent of being burned at the stake." All discussion of the Eighth Amendment in relation to the death penalty turns on the time (that is, duration) of an execution. See David Wills, *Killing Times: The Temporal Technology of the Death Penalty* (New York: Fordham University Press, 2019), esp. 17–53.
4. Wills, *Killing Times,* 214.
5. G. W. F. Hegel, *Hegel's Phenomenology of Spirit,* trans. A. V. Miller (Oxford: Oxford University Press, 1977), 100. The discussion that follows is developed in considerably more detail in Wills, *Killing Times,* 101–14.
6. Hegel, *Hegel's Phenomenology of Spirit,* 489.
7. G. W. F. Hegel, *Hegel's Philosophy of Nature,* trans. A. V. Miller (Oxford: Oxford University Press, 2007), 443.
8. Hegel, *Hegel's Philosophy of Nature,* 360.
9. Hegel, *Hegel's Philosophy of Nature,* 368.
10. Hegel, *Hegel's Philosophy of Nature,* 369, 368.
11. Hegel, *Hegel's Philosophy of Nature,* 381.
12. Hegel, *Hegel's Philosophy of Nature,* 366.

13. Hegel, *Hegel's Philosophy of Nature*, 367, 397.
14. Hegel, *Hegel's Philosophy of Nature*, 392.
15. Hegel, *Hegel's Philosophy of Nature*, 366.
16. Derrida, *Death Penalty*, 1:257.
17. Derrida, *The Death Penalty*, vol. 2, trans. Elizabeth Rottenberg (Chicago: University of Chicago Press, 2017), 2:199–200.
18. Nick Bostrom, "Ethical Issues in Advanced Artificial Intelligence," in *Science Fiction and Philosophy: From Time Travel to Superintelligence*, ed. Susan Schneider (Oxford: Wiley-Blackwell, 2009), 282.
19. Yuval Noah Harari, *Sapiens: A Brief History of Humankind* (New York: Harper, 2015), 403.
20. Ken Klippenstein and Matthew Petit, "Biden Admin Deployed Air Force Team to Israel to Assist with Targets, Document Suggests," *Intercept*, January 11, 2024, https://theintercept.com/2024/01/11/israel-air-force-targeting-intelligence/.
21. Blanchot, *Instant*, 9, 5.

Preface and Acknowledgments

1. Jacques Derrida and Jean Birnbaum, *Learning to Live Finally: The Last Interview*, trans. Pascale-Anne Brault and Michael Naas (New York: Palgrave Macmillan, 2007), 32–33.
2. Jacques Derrida, *The Beast and the Sovereign*, vol. 2, trans. Geoffrey Bennington, ed. Michel Lisse, Marie-Louise Mallet, and Ginette Michaud (Chicago: University of Chicago Press, 2011), 2:132; and Jacques Derrida, *Séminaire: La bête et le souverain, Volume 2 (2002–2003)* (Paris: Galilée, 2010), 195.
3. On the link between immortality and sovereignty, see Jacques Derrida, *The Beast and the Sovereign*, vol. 1, trans. Geoffrey Bennington (Chicago: University of Chicago Press, 2009), 1:42–43. On immortality and the death penalty, see Jacques Derrida, *The Death Penalty*, vol. 1, trans. Peggy Kamuf (Chicago: University of Chicago Press, 2013), 255–56. On immortality and autoaffection, see Jacques Derrida, *Speech and Phenomena*, trans. David B. Allison (Evanston, Ill.: Northwestern University Press, 1973), 78–79. On immortality, infinite presence, and God, see the following works by Derrida: *Of Grammatology*, trans. Gayatri Chakravorty Spivak (Baltimore, Md.: Johns Hopkins University Press, 1976), 70–71, a translation of *De la grammatologie* (Paris: Les Éditions de Minuit, 1967); and "The Night Watch," in *Derrida and Joyce*, ed. Andrew J. Mitchell and Sam Slote, trans. Pascale-Anne Brault and Michael Naas (Albany: State University of New York Press, 2012), 99–101.

4. See, e.g., Maurice Blanchot and Jacques Derrida, *The Instant of My Death/Demeure: Fiction and Testimony* (Stanford, Calif.: Stanford University Press, 2000), 2–11; and Martin Heidegger, *The Fundamental Concepts of Metaphysics: World, Finitude, Solitude,* trans. William McNeill and Nicholas Walker (Bloomington: Indiana University Press, 1995). References to the German are taken from Martin Heidegger, *Gesamtausgabe II. Abteilung: Vorlesungen 1923–1924. Band 29/30 Die Grundbegriffe der Metaphysik: Welt—Endlichkeit—Einsamkeit* (Frankfurt am Main: Vittorio Klostermann, 1992).
5. For different versions of this argument, see the following works by Hägglund and Baudrillard: Martin Hägglund, *Radical Atheism* (Stanford, Calif.: Stanford University Press, 2008); Martin Hägglund, *Dying for Time* (Cambridge, Mass.: Harvard University Press, 2012); Martin Hägglund, *This Life* (New York: Pantheon, 2019); Jean Baudrillard, *The Vitalist Illusion,* ed. Julia Witwer (New York: Columbia University Press, 2000); and Jean Baudrillard, *Impossible Exchange* (London: Verso, 2001).
6. In addition to Hägglund and Baudrillard, prominent vocal opponents of immortality and transhumanism include Leon Kass, Francis Fukuyama, Jürgen Habermas, and N. Katherine Hayles. See Leon Kass, *Life, Liberty and the Defense of Dignity* (San Francisco, Calif.: Encounter, 2002); Francis Fukuyama, *Our Posthuman Future* (New York: Farrar, Straus & Giroux, 2002); Jürgen Habermas, *The Future of Human Nature* (Cambridge: Polity, 2003); and N. Katherine Hayles, *How We Became Posthuman* (Chicago: University of Chicago Press, 1999).
7. Derrida, *Of Grammatology,* 68. On the problem of finitude in Derrida's work, as well as his troubling of it, see Rodolphe Gasché, *The Tain of the Mirror* (Cambridge, Mass.: Harvard University Press, 1988); Geoffrey Bennington, "Derridabase," in *Jacques Derrida,* trans. Geoffrey Bennington (Chicago: University of Chicago Press, 1993), 115, 117; and Kas Saghafi, "The Desire for Survival," in *Desire in Ashes,* ed. Simon Morgan Wortham and Chiara Alfano (London: Bloomsbury, 2015), 139–60.
8. Derrida, *Learning to Live Finally,* 25. Just before his death, Derrida left a brief note that he wished to be read at his funeral. It was published in 2005 as his "Derniers mots," and then in 2007 in English translation as "Final Words." In it, he repeats the affirmation of survival while maintaining a certain indecision concerning the possibility of life after death: "Smile for me, he says, as I will have smiled for you until the end. Always prefer life and constantly affirm survival. . . . I love you and am smiling at you from wherever I am." Jacques Derrida, "Final Words," trans. Gila Walker, *Critical Inquiry* 33, no. 2 (2007): 462.
9. Derrida, *Beast,* 2:124.

Introduction

1. Derrida, *Of Grammatology,* 313. As Wills puts it, "The prosthetic or prostheticized body is one whose naturality was never intact, never without preparation, organization, articulation, and appeal for its artificial necessities." David Wills, *Inanimation: Theories of Inorganic Life* (Minneapolis: University of Minnesota Press, 2016), xii.
2. Although transhumanism is generally considered to be a subset of posthumanism, there is a marked difference between this stream of thinking and that represented by the critical posthumanisms of Wolfe, Hayles, Claire Colebrook, Stefan Herbrechter, and many others, for example in the Critical Posthumanism Network, who are suspicious of transhumanist thought. For an excellent overview of the recent literature of posthumanism, see Nicole Anderson, "Animal–Human Difference: The Deconstructive Force of Posthumanism," *Studies in Literature, Culture, and the Environment* 11 (2023): 211–25. My criticisms of the posthuman in what follows are of its transhumanist inclinations rather than of this critical tradition.
3. On the Singularity, see Ray Kurzweil, *The Singularity Is Near: When Humans Transcend Biology* (New York: Penguin, 2006); and Vernor Vinge, "The Coming Technological Singularity," *Whole Earth Review* (1993), https://accelerating.org/articles/comingtechsingularity. On the 2045 Initiative, see 2045 Strategic Social Initiative, http://www.2045.com/; and BBC Two, "The Immortalist," *Horizon,* March 15, 2016, https://www.bbc.co.uk/programmes/p03mtthg. On the elimination of the biological necessity of death, see Aubrey de Grey and Michael Rae, *Ending Aging* (New York: St. Martin's Griffin, 2008); and Ray Kurzweil and Terry Grossman, *Transcend: Nine Steps to Living Well Forever* (New York: Rodale, 2009). On superintelligence, see Nick Bostrom, *Superintelligence* (Oxford: Oxford University Press, 2014); and Yuval Noah Harari, *Homo Deus* (New York: Penguin, 2016).
4. Heinrich von Kleist, "On the Marionette Theater," trans. Christian-Albrecht Gollub, in *German Romantic Criticism,* ed. A. Leslie Willson (New York: Continuum, 1982), 244.
5. It has become a cliché to point out the complicity between religious and transhumanist discourse on the one hand and transhumanism and humanism on the other. On the link between transhumanism and religion, see, e.g., Ronald Cole-Turner, ed., *Transhumanism and Transcendence* (Washington, D.C.: Georgetown University Press, 2011); Jacob Shatzer, *Transhumanism and the Image of God* (Downers Grove, Ill.: InterVarsity Press, 2019); Michael Hauskeller, *Mythologies of Transhumanism* (Cham,

Switzerland: Palgrave Macmillan, 2016); and Mark O'Connell, *To Be a Machine* (New York: Doubleday, 2017). On the humanism of the posthuman and transhuman, see Hayles, *How We Became Posthuman;* and Cary Wolfe, *What Is Posthumanism?* (Minneapolis: University of Minnesota Press, 2009). For an account of the utopian rhetoric of contemporary transhuman thought, see Andrew Pilsch, *Transhumanism* (Minneapolis: University of Minnesota Press, 2017). The present study differs from many of these approaches in that although the upshot of many of these accounts has tended to be rejection of transhumanism, I argue that the problem to which transhumanism itself responds is irreducible and thus calls for another mode of analysis.

6. Three representative essays by Bostrom on the problem of radical life extension are: Nick Bostrom, "The Case against Aging" (2000), https://nickbostrom.com/aging/aging; Nick Bostrom, "The Fable of the Dragon-Tyrant," *Journal of Medical Ethics* 31, no. 5 (2005): 273–77; and Nick Bostrom with Rebecca Roache, "Ethical Issues in Human Enhancement," in *New Waves in Applied Ethics,* ed. Jesper Ryberg, Thomas Petersen, and Clark Wolf (London: Palgrave Macmillan, 2009), 120–52. Bostrom is exemplary of current trends in transhumanism—not only because he is one of the founders of the movement, along with David Pearce, but also because he presents the strongest and most philosophically rigorous version of it. For other representatives, see Max More, "Transhumanism: Towards a Futurist Philosophy" (1990), www.maxmore.com/transhum.htm; and John Harris, *Enhancing Evolution* (Princeton, N.J.: Princeton University Press, 2007). For other proponents of finitude, either on ethical or humanist grounds, see Kass, *Life, Liberty;* Fukuyama, *Our Posthuman Future;* and Habermas, *Future of Human Nature.*
7. For an important analysis of the discourse of extinction and its relation to deextinction efforts, see Joshua Schuster, *What Is Extinction? A Natural and Cultural History of Last Animals* (New York: Fordham, 2023).
8. For a study of the contemporary effects of genomics and informatics on conceptions of life, see Eugene Thacker, *The Global Genome* (Cambridge, Mass.: MIT Press, 2005), esp. 1–130; and Robert Mitchell and Phillip Thurtle, eds., *Data Made Flesh* (New York: Routledge, 2004). For an account of the futural structure of the grammar of gene research, see Michael Fortun, "The Human Genome Project: Past, Present, and Future Anterior," in *Science, History and Social Concern,* ed. Garland E. Allan and Roy MacLeod (Dordrecht, Netherlands: Kluwer, 2002), 44–49; and Michael Fortun, *Promising Genomics* (Berkeley, Calif.: University of California Press, 2008).

9. I am indebted on this point to Naas's definitive articulation of the problematic. See Michael Naas, *Derrida from Now On* (New York: Fordham University Press, 2008); and Michael Naas, "The Mother of All Phantasms," in Mitchell and Slote, *Derrida and Joyce,* 163–98.
10. This phrase was evidently popularized by *Time* magazine in their 2013 cover story on Google's Calico (California Life Company), entitled "Can Google Solve Death?"
11. On the indefatigable, see especially Derrida, "Night Watch."
12. In "Thoughts for the Times on War and Death," Freud famously declares, "It is indeed impossible to imagine our own death . . . hence the psychoanalytic school . . . assertion that at bottom no one believes in his own death, or, to put the same thing in another way, that in the unconscious every one of us is convinced of his own immortality." Sigmund Freud, *The Standard Edition of the Complete Psychological Works of Sigmund Freud, Volume 14,* ed. and trans. James Strachey (London: Hogarth Press and the Institute of Psycho-analysis, 1957), 289; see also 296, 299. On the problem of immortality in psychoanalysis, see Adrian Johnston, "Life Terminable and Interminable: The Undead and the Afterlife of the Afterlife—A Friendly Disagreement with Martin Hägglund," *New Centennial Review* 9, no. 1 (2009): 147–89, and Alla Ivanchikova, "The Fantasy of Technoimmortality and the Psychoanalytic Infinite," *Comparatist* 45 (2021): 64–89.
13. On this phrase, see Geoffrey Bennington, *No Half No End* (Edinburgh: Edinburgh University Press, 2011), 65–85.
14. For Derrida's own call to think *différance* beyond this difference, see Derrida, *Speech and Phenomena,* 87. On the irreducibility of the phantasm of immortality within the logic of autoimmunity, see Jacques Derrida, "The Deconstruction of Actuality," in *Negotiations: Interventions and Interviews, 1971–2001,* trans. Elizabeth Rottenberg (Stanford, Calif.: Stanford University Press, 2002), 102–3.
15. Naas voices a similar sentiment: "In this respect the phantasm will be something like the metaphysical shadow of Derrida's concept of *living on,* one of the reasons, I would suggest, why Derrida retained it, since by never shaking completely free of its metaphysical double Derridean living on or survival exerts a *critical force* against this powerful metaphysical phantasm. While the phantasm promises a life beyond life, a pure life without technique or the machine, a life beyond history and time, the concept of living on or of survival acknowledges history and convention and promises an intensification of life, a life in death, but never any immortality." Naas, *Derrida from Now On,* 201. See also Wills, *Inanimation,* 97.

16. For an attempt to do just this by way of the question of finitude, see Michael Haworth, *Neurotechnology and the End of Finitude* (Minneapolis: University of Minnesota Press, 2018).
17. These two uses roughly correspond to the two senses that Derrida identifies in the word "survival," specifically as a translation of the Benjaminian opposition between *Überleben* and *Fortleben*. See Jacques Derrida, *The Ear of the Other*, ed. Christie McDonald, trans. Peggy Kamuf (New York: Schocken, 1985), 121–24. I discuss this link, as well as its particular importance for prosthetic immortalization, in Adam R. Rosenthal, "Love of Life: Deconstruction, Biotech and the Survival of Indefinite Life," *Oxford Literary Review* 40, no. 2 (2018): 156–80.

1. Life as Generation in Plato

1. Plato, *Lysis, Symposium*, and *Gorgias*, trans. W. R. M. Lamb (Cambridge, Mass.: Harvard University Press, 2001), 193.
2. Plato, *Symposium* (Lamb), 195. The Greek reads: "Εἰ τοίνυν, ἔφη, πιστεύεις ἐκείνου εἶναι φύσει τὸν ἔρωτα, οὗ πολλάκις ὡμολογήκαμεν, μὴ θαύμαζε. ἐνταῦθα γὰρ τὸν αὐτὸν ἐκείνῳ λόγον ἡ θνητὴ φύσις ζητεῖ κατὰ τὸ δυνατὸν ἀεὶ τὸ εἶναι ἀθάνατος. δύναται δὲ ταύτῃ μόνον, τῇ γενέσει, ὅτι ἀεὶ καταλείπει ἕτερον νέον ἀντὶ τοῦ παλαιοῦ" (194).
3. Plato, *Symposium* (Lamb), 197. The Greek reads: "τούτῳ γὰρ τῷ τρόπῳ πᾶν τὸ θνητὸν σῴζεται, οὐ τῷ παντάπασι τὸ αὐτὸ ἀεὶ Βεἶναι ὥσπερ τὸ θεῖον, ἀλλὰ τῷ τὸ ἀπιὸν καὶ παλαιούμενον ἕτερον νέον ἐγκαταλείπειν οἷον αὐτὸ ἦν" (196).
4. Plato, *Symposium* (Lamb), 191. The Greek reads: "ἡ γὰρ ἀνδρὸς καὶ γυναικὸς συνουσία τόκος ἐστίν. ἔστι δὲ τοῦτο θεῖον τὸ πρᾶγμα, καὶ τοῦτο ἐν θνητῷ ὄντι τῷ ζῴῳ ἀθάνατον ἔνεστιν, ἡ κύησις καὶ ἡ γέννησις. τὰ δ' ἐν τῷ ἀναρμόστῳ ἀδύνατον Dγενέσθαι. ἀνάρμοστον δ' ἐστὶ τὸ αἰσχρὸν παντὶ τῷ θείῳ, τὸ δὲ καλὸν ἁρμόττον" (190).
5. In classical Greek, κύησις, "engendering" or "conception," along with the older verb form κυέω, and later κύω, "to conceive," "to be pregnant," are generally only said of female subjects. Τόκος, in contrast, may refer either to the "childbirth" and "parturition" of women or to gender-neutral senses of "offspring" and "production," just as the verb τίκτω, "to bring into the world, to engender, to beget, to produce," can be used with both genders. In this passage, Diotima ascribes the typically feminine attribute of pregnancy, κύησις, indiscriminately to both genders, so as to relate it, without essential difference, both to the process of begetting more generally, in τόκος, and to the "producing, creating" activity of γέννησις, and

γεννάω, "to produce, to bring into being, to create," said of both natural and nonnatural processes. Γενέσει, or γένεσις, "origin, source, production, generation, coming into being, creation," is similarly used throughout the speech to refer to this basic attribute of productivity, a term at work in "natural" and "nonnatural" births alike. Throughout this chapter, I have given "generation" as the name for the general form of (re)productivity at stake in Diotima's speech, which is supposed to underly the relationship between biological procreation on the one hand and productivity as poiesis on the other (206b–c; 208e–212a). On Plato's use—and abuse—of these highly charged terms in the dialogue, see A. W. Price, *Love and Friendship in Plato and Aristotle* (Oxford: Oxford University Press, 1989), 17–18, 26–28; and Plato, *Symposium,* ed. Kenneth Dover (Cambridge: Cambridge University Press, 1980), 147.

6. Plato, *Symposium* (Lamb), 193.
7. Though I cannot pursue the connection here, there is a case to be made for generalizing Diotima's view of life in the *Symposium,* for Plato's thought in general. For the most salient articulation of such a connection, see Michael Naas, *Plato and the Invention of Life* (New York: Fordham University Press, 2018), where he identifies the essential problem of life for Plato through its drive for a beyond of mere or bare mortal living. In particular, Naas argues for the silent inscription of a "*real life* or *life itself*" running throughout Plato's dialogues (14). Such an "essence or form of life" would be opposed, for Plato, to "what is *called* life" while serving to orient it in the direction of its transformation. Thus, although Naas does not read the *Symposium* in connection to this movement, focusing instead on the *Statesman,* at stake for him is nonetheless a question similar to the one raised here: the necessity for Plato of understanding mortal life as being essentially driven to immortality.
8. Diotima, of course, does not speak directly in the *Symposium.* She speaks through Socrates, who, at the moment of narration of Plato's dialogue, speaks through Apollodorus, who himself had recently conveyed the narrative to Glaucon, and is presently doing so to his unnamed companion, because it had already been recounted to him by Aristodemus, who was himself in attendance. For an excellent account of the debate surrounding Diotima's role in the *Symposium,* and the likelihood that she names a "real" woman by this or another name (most likely Aspasia), see David Halperin, "Why Is Diotima a Woman?," in *A Hundred Years of Homosexuality* (New York: Routledge, 1989), 113–51.
9. Diotima leaves this point ambiguous by saying "εἴπερ": "So when he has begotten a true virtue and has reared it up he is destined to win

the friendship of Heaven; he, above all men, is immortal" (τεκόντι δὲ ἀρετὴν ἀληθῆ καὶ θρεψαμένῳ ὑπάρχει θεοφιλεῖ γενέσθαι, καὶ εἴπερ τῳ ἄλλῳ ἀνθρώπων ἀθανάτῳ καὶ ἐκείνῳ) (212a). Gill translates the passage as follows: "It's someone who's given birth to true virtue and brought it up who has the chance of becoming loved by the gods, and immortal—if any human being can be immortal" (50). See Plato, *Symposium*, trans. Christopher Gill (New York: Penguin, 2003).

10. In distinction to the *Phaedo, Meno,* and *Phaedrus,* nowhere in the *Symposium* does Plato explicitly state that the soul is immortal. On this problem, see Plato, *Symposium* (Dover), 149–50.
11. On this point, see Plato, *Symposium* (Lamb), 191c. As Nussbaum points out, the wish of the halved men of Aristophanes's myth is necessarily twofold. On the one hand, it is to conjoin each bodily half to form a whole. On the other hand, it is to merge their souls with these newly formed bodies. See Martha Nussbaum, *The Fragility of Goodness* (Cambridge: Cambridge University Press, 1986), 175–76.
12. The figures of wholeness that are literalized in Aristophanes's myth (τοῦ ὅλου οὖν τῇ ἐπιθυμίᾳ καὶ διώξει ἔρως ὄνομα [192e–193a]) become radicalized in Diotima's. Immortality remains conceivable only on the model of these figures, but now as absolutely whole, rather than whole in one attribute alone. What is sought is completeness of soul and body alike, which requires conceiving of completeness on the model of the forms rather than on that of the mere body, which would be derivative with respect to them.
13. Plato, *Symposium* (Lamb), 195.
14. Plato, *Symposium* (Lamb), 197. The Greek reads: "μὴ οὖν θαύμαζε εἰ τὸ αὑτοῦ ἀποβλάστημα φύσει πᾶν τιμᾷ: ἀθανασίας γὰρ χάριν παντὶ αὕτη ἡ σπουδὴ καὶ ὁ ἔρως ἕπεται" (196).
15. Plato, *Symposium* (Lamb), 191. The Greek reads: "κυοῦσιν γάρ, ἔφη, ὦ Σώκρατες, πάντες ἄνθρωποι καὶ κατὰ τὸ σῶμα καὶ κατὰ τὴν ψυχήν, καὶ ἐπειδὰν ἔν τινι ἡλικίᾳ γένωνται, τίκτειν ἐπιθυμεῖ ἡμῶν ἡ φύσις" (190). Compare to the following passage from the *Laws* (Bury's translation): "A man shall marry when he is thirty years old and under thirty-five, bearing in mind that this is the way by which the human race, by nature's ordinance, shares in immortality, a thing for which nature has implanted in everyone a keen desire. The desire to win glory, instead of lying in a nameless grave, aims at a like object. Thus mankind is by nature coeval with the whole of time, in that it accompanies it continually both now and in the future; and the means by which it is immortal is this:—by leaving behind it children's children and continuing ever one and the

same, it thus by reproduction shares in immortality" (721b–c). Plato, *Laws, Volume 1: Books 1–6,* trans. R. G. Bury (Cambridge, Mass.: Harvard University Press, 1926), 311–13.

16. The primary references for Diotima are human and animal life. Elsewhere, however, in the *Phaedo,* plant life is also included within the category of that which "is born" or "generated": " 'Well then, don't look at this,' he said, 'only from the human angle, if you want to understand it more easily, but from that of all animals and plants, and by looking collectively at all things that come into being [ἔχει γένεσιν] let's see whether everything comes into being in this way, from nowhere but opposites from their opposite' " (70e). Plato, *Euthyphro, Apology, Crito,* and *Phaedo,* ed. and trans. Christopher Emlyn-Jones and William Preddy (Cambridge, Mass.: Harvard University Press, 2017), 342–45. In the *Timaeus,* plants, like animals, are said to "partake of life" and thus to "justly and with perfect truth be termed a living creature" (77b). Moreover, plant life is said to possess a third kind of soul, alongside those of humans and animals, and to that end also to bear desires (77b). Plato, *Timaeus, Critias, Cleitophon, Menexenus,* and *Epistles,* trans. R. G. Bury (Cambridge, Mass.: Harvard University Press, 1929), 202–5. On the relation between each of these forms of life in Plato, see especially Naas, *Plato,* 30–34. As Naas points out, plants would seem to be alive for Plato, but "they do not seem to have a *bios,*" in the sense of being capable of good or bad lives, as animals and humans are (239n8).
17. This paradoxical relation between mortality and immortality, at work within all life-forms, is first modeled by Eros himself in Diotima's account, when she teaches Socrates about his "in-between" character and establishes that he is a *daimon* and thus neither mortal nor immortal (203a).
18. This logic is laid out explicitly: "Every mortal thing is preserved in this way; not by keeping it exactly the same for ever, like the divine, but by replacing what goes off or is antiquated with something fresh, in the semblance of the original. Through this device, Socrates, a mortal thing partakes of immortality, both in its body and in all other respects; by no other means can it be done. So do not wonder if everything naturally values its own offshoot; since all are beset by this eagerness and this love with a view to immortality," Plato, *Symposium,* (Lamb), 208a–b, 197.
19. On the coherences of this vision of love, which aims at its own dissolution, with the eros of Aristophanes, see Nussbaum, *Fragility of Goodness,* 175–76. For Nussbaum, Aristophanes's is already a desire to end all desires, an unrealizable desire for a totality without desire. For a similar

argument, one concerning the logic at work within the conception of immortality in the *Timaeus,* see Jacques Derrida, "Plato's Pharmacy," in *Dissemination,* trans. Barbara Johnson (Chicago: University of Chicago Press, 1981), 61–172, 101–2.

20. Plato, *Symposium* (Lamb), 193. The Greek reads: "ἔστι γάρ, ὦ Σώκρατες, ἔφη, οὐ τοῦ καλοῦ ὁ ἔρως, ὡς σὺ οἴει. Ἀλλὰ τί μήν; Τῆς γεννήσεως καὶ τοῦ τόκου ἐν τῷ καλῷ. Εἶεν, ἦν δ' ἐγώ. Πάνυ μὲν οὖν, ἔφη. τί δὴ οὖν τῆς γεννήσεως; ὅτι ἀειγενές ἐστι καὶ ἀθάνατον ὡς θνητῷ ἡ γέννησις. ἀθανασίας δὲ ἀναγκαῖον ἐπιθυμεῖν μετὰ ἀγαθοῦ ἐκ τῶν ὡμολογημένων, εἴπερ τοῦ ἀγαθὸν ἑαυτῷ εἶναι ἀεὶ ἔρως ἐστίν. ἀναγκαῖον δὴ ἐκ τούτου τοῦ λόγου καὶ τῆς ἀθανασίας τὸν ἔρωτα εἶναι" (192). This is the second time Diotima expresses this sentiment. See also 206a.
21. For the relationship between Diotima, Socrates, and Plato, see Halperin, "Why Is Diotima a Woman?," 125–27. Vlastos locates Plato's break with Socrates at 206b, when he admits that he cannot answer her question. See Gregory Vlastos, "The Individual as Object of Love in Plato," in *Platonic Studies* (Princeton, N.J.: Princeton University Press, 1981), 3–34, 21n58.
22. Vlastos, "Individual as Object of Love," argues that this is the first time in Greek literature that one finds a figure of male pregnancy (21n59). On the novelty of Plato's notion of male eros as creative and procreative, see also Halperin, "Why Is Diotima a Woman?," 137–38, 153.
23. As was clarified above initially at 206e–207a, Diotima explains that the aim of mortal life is the possession of the good for oneself, by way of immortality. Later, in the culminating moments of Diotima's speech, at 210e–212a, rather than a means to this end, immortality becomes the end itself, which the ideal form of beauty exemplifies, being "ever-existent" (210e), "eternal and unitary" (211b), and "divine" (211e).
24. Plato, *Symposium* (Lamb), 194–95.
25. Plato, *Symposium* (Lamb), 197. The Greek reads: "τούτῳ γὰρ τῷ τρόπῳ πᾶν τὸ θνητὸν σῴζεται, οὐ τῷ παντάπασι τὸ αὐτὸ ἀεὶ Βεῖναι ὥσπερ τὸ θεῖον, ἀλλὰ τῷ τὸ ἀπιὸν καὶ παλαιούμενον ἕτερον νέον ἐγκαταλείπειν οἷον αὐτὸ ἦν. ταύτη τῇ μηχανῇ, ὦ Σώκρατες, ἔφη, θνητὸν ἀθανασίας μετέχει, καὶ σῶμα καὶ τἆλλα πάντα· ἀδύνατον δὲ ἄλλῃ. μὴ οὖν θαύμαζε εἰ τὸ αὑτοῦ ἀποβλάστημα φύσει πᾶν τιμᾷ· ἀθανασίας γὰρ χάριν παντὶ αὕτη ἡ σπουδὴ καὶ ὁ ἔρως ἕπεται" (196).
26. In Plato, *Symposium* (Lamb): Beauty, before it becomes the ideal form of beauty, is what one loves, and one loves it because it is good, and one loves the good because the good makes one happy (204d–206b). Love is of having the good forever. Subsequently, beauty becomes that in which

one can reproduce (and reproduction in body and in mind are ways that mortals partake in immortality). It is the vehicle of that action (206c–d). The object of love, Diotima says outright, is not beauty, but reproduction *in the beautiful* (206e). To say, as I have, that Plato's notion of reproduction in the beautiful extends to all living beings is not to say that there is no other of beauty—above all for Plato. Rather, it is to say that even being with the nonbeautiful (or the ugly)—which does not stimulate self-reproduction in the way that the beautiful does—is itself a mode of being with respect to beauty, rather than a wholly different mode of being. To be with the non-beautiful is to be less reproductive, less stimulated to reproduction—but it is not, if we follow Diotima's logic, to be in a mode of nonreproduction. Where would one even find the ground for such a mode?

27. What if nonhuman animals reproduced not only through generation but also art-making practices? For an excellent discussion of the history and intellectual stakes of this question, see Deirdre Madeleine Smith, "The Origins of Animal Art," *Configurations* 31 (2023): 61–90.
28. Plato, *Symposium* (Lamb), 197.
29. To put it in Aristotelian causal terms, we could say that on Plato's account, while the final and efficient causes of man's life are definite (immortality and eros, respectively), the material and formal causes are essentially indefinite. By contrast, (so the thinking goes), the bird's life is definite through and through, insofar as all its love is poured into making other birds.
30. Though no concept of reproduction, in the current, biological sense, could possibly apply, what Plato articulates, by way of Diotima, is simply the logic of reproduction, as Derrida identifies; see Jacques Derrida, *Life Death,* trans. Pascale-Anne Brault and Michael Naas, ed. Peggy Kamuf (Chicago: University of Chicago Press, 2020).
31. On the self-interested nature of Platonic love, see especially Vlastos, "Individual as Object of Love," which contrasts Platonic ἔρως with Aristotelian φιλία.
32. For a reading of Diotima's speech that emphasizes fecundity, through "regrowth," "becoming in time," and "permanent generation and regeneration," over and above the narcissistic bent of autoextension and immortality, see Luce Irigaray, "Sorcerer Love: A Reading of Plato's *Symposium,* Diotima's Speech," *Hypatia* 3, no. 3 (1989): 32–44, 39.
33. On the far-reaching scope of life encompassed by this vision, see especially 207: "'What do you suppose, Socrates, to be the cause of this love and desire? For you must have observed the strange state into which all the animals [πάντα τὰ θηρία] are thrown, whether going on

earth or winging the air [καὶ τὰ πεζὰ καὶ τὰ πτηνά], when they desire to beget: they are all sick and amorously disposed. . . . As for men,' she said, 'one might suppose they do these things on the promptings of reason; but what is the cause of this amorous condition in the animals [τὰ δὲ θηρία]?' " (207a–c).

34. Reproductive futurism is Edelman's figure. See Lee Edelman, *No Future* (Durham, N.C.: Duke University Press, 2004).
35. In the *Phaedo,* 100c–e, Socrates makes use of the same language of participation to speak of how beautiful things "partake" of the beautiful. He also acknowledges that although he is sure that this is how it happens, he is ignorant of the precise mechanism by which it does. On this, see Plato, *Symposium* (Dover), 6–7. Does not the legibility of reproduction as reproduction require a notion of species, or some other unifying concept, through which one thinks to associate separate entities as part of the same? This is no simple task, as even anecdotal zoological evidence shows. On the animal kingdom's ignorance of the relation between sex and propagation, see Daniel Milo, *Good Enough* (Cambridge, Mass.: Harvard University Press, 2019), 210–26. See also François Jacob, *The Logic of Life,* trans. Betty E. Spillman (Princeton, N.J.: Princeton University Press, 1973), on the historical resistance to associating acts of sex with the generative power of producing children. Jonas articulates a parallel argument: Hans Jonas, *Phenomenon of Life* (Evanston, Ill.: Northwestern University Press, 2001). Theorizing the meaning of life from a purely physicalist point of view, Jonas contends that for a "mathematical God," deprived of "the benefit of organic experience," "life itself would not even become visible, nor yet a blank in its place. Eye would not be recognized as eye, feeler not as feeler, organism not as organism" (87). For Jonas, in sum, it is the inwardness of experience constitutive of life that itself provides the possibility of "seeing" life as that which is self-directed, and consequently irreducible to or untranslatable into the language of physical systems. Thus, while for Plato it is a universal participation in divine immortality that allows for the experience of reproduction in all of its variegated forms, for Jonas it is the inwardness of experience as such that allows for the perception of life.
36. Compare this to what Naas identifies as the general structure of life in Plato's works: "We will thus see that the life that Plato privileges in several of his most important later dialogues is, in the end, not bare life . . . but something like *real life* or *life itself*—that is, a life beyond or in excess of what we call life. By thinking being (whether in the form of the Forms or the immortal soul) in terms of life, Plato is ultimately able to discover

(or, as I will argue, to *invent*) this notion of *life itself*—that is, a *real* or *true* life that would be opposed to all merely *biological* or *animal* life, a form of life, then, that would be more valuable than everything we call life and every life that can actually be lived." Naas, *Plato,* 4.

2. Derrida and the Mortal Immortality of Sovereignty

1. Derrida, *Beast,* 1:43.
2. Compare to these remarks by Derrida: "But since this happens all the time, pure sovereignty does not exist; it is always in the process of positing itself by refuting itself, by denying or disavowing itself; it is always in the process of autoimmunizing itself, of betraying itself by betraying the democracy that nonetheless can never do without it," and "more precisely, since it never succeeds in doing this except in a critical, precarious, and unstable fashion, sovereignty can only *tend,* for a limited time, to reign without sharing. It can only tend toward imperial hegemony." Jacques Derrida, *Rogues: Two Essays on Reason,* trans. Pascale-Anne Brault and Michael Naas (Stanford, Calif.: Stanford University Press, 2005), 101, 102.
3. Derrida, *Beast/Bête,* 1:42/71–72. The French reads: "S'il y a une structure prothétique du *Léviathan* comme animal ou monster politique, cela tient à sa structure conventionnelle, thétique, contractuelle. L'opposition entre *physis* et *nomos* (nature et loi), comme opposition entre *physis* et *thesis* (nature et convention, ou nature et position) joue ici à plein et de façon décisive. Il s'ensuit que la loi, la souveraineté, l'institution de l'État sont historiques et toujours provisoires, disons déconstructibles, par essence fragiles ou finies ou mortelles, même si la souveraineté est *posée comme immortelle.* Elle est posée comme immortelle et indivisible précisément parce qu'elle est mortelle, et divisible, le contrat ou la convention étant destinés à lui assurer ce qu'elle n'a pas ou n'est pas naturellement."
4. Derrida, *Beast/Bête,* 1:40/69.
5. Thomas Hobbes, *Leviathan,* ed. Edwin Curley (Indianapolis: Hackett, 1994), 144–45.
6. Derrida, *Beast/Bête,* 1:43/73. The French reads: "Dans le paragraphe du *Léviathan* que je vais lire, je souligne le lexique de la *protection,* là où ce lexique et sa logique expliquent le paradoxe de la mortelle immortalité de la souveraineté. La souveraineté, âme et donc vie de l'État, respiration artificielle de l'État, n'est posée, instituée, promise, contractée, *artificiellement, comme immortelle* que parce qu'elle est *naturellement mortelle.* C'est

la technique prothétique et artificielle qui l'immortalise ou lui assure en tout cas une survie indéfinie."

7. On the perfectibility of democracy, see, e.g., Derrida, *Rogues,* 90–91, 150–51; and Jacques Derrida, *Politiques de l'amitié: Suivi de L'oreille de Heidegger* (Paris: Galilée, 1994), 128–29.
8. Readers of Derrida, *Beast,* vol. 2, will pick up on a certain disparity between Derrida's problematization of the "as such" of mortality there and what I am here calling "mortality as such" to speak of the inveterate divisibility of sovereignty. In this way, I distinguish between mortality as divisibility (of structure) and mortality as the (ontological) privilege of Dasein in being-toward-death.
9. Derrida, *Beast,* 1:66.
10. Derrida, *Beast,* 1:66–67.
11. Derrida, *Beast,* 1:67.
12. Derrida glosses *ipseity* in the following manner: "It seems difficult to think such a desire for or naming of democratic space without the rotary motion of some quasi-circular return or rotation toward the self, toward the origin itself, toward and upon the self of the origin, whenever it is a question, for example, of sovereign self-determination, of the autonomy of the self, of the *ipse,* namely, of the one-self that gives itself its own law, of autofinality, autoteley, self-relation as being in view of the self, beginning by the self with the end of self in view—so many figures and movements that I will call from now on, to save time and speak quickly, to speak in round terms, *ipseity* in general. By *ipseity* I thus wish to suggest some 'I can,' or at the very least the power that *gives itself* its own law, its force of law, its self-representation, the sovereign and reappropriating gathering of self in the simultaneity of an assemblage or assembly, being together, or 'living together,' as we say." Derrida, *Rogues,* 10–11.
13. Derrida repeats, on several occasions, that there is no simple nonsovereign opposition to sovereignty. However, the purpose of the seminar is precisely to articulate another logic, one that no longer simply remains within the confines of this opposition. See, e.g., Derrida, *Beast,* 1:289–91, 1:301–2. In "The University without Condition," it is the university that is appealed to, as requiring the articulation of different relations of "weak force," so as not to succumb to this tendency. Jacques Derrida, "The Future of the Profession, or The University without Condition," Presidential Lecture at the University of California, April 15, 1999.
14. On the ubiquity of sovereign identity, from self to nation-state, see also Michael Naas, "'One Nation . . . Indivisible': Jacques Derrida on the Autoimmunity of Democracy and the Sovereignty of God," in *Research in*

Phenomenology 36 (2006): 15–44. Naas helpfully recalls Derrida's emphasis, linking the "I" of the "I pledge" to a "kind of originary fidelity before the bond of any nation-state or religion, a faith that opens up the nation-state and every national context to something that exceeds it" (16).

15. Ernst Kantorowicz, *The King's Two Bodies* (Princeton, N.J.: Princeton University Press, 1957), 5.
16. Kantorowicz, *King's Two Bodies,* 45.
17. Kantorowicz, *King's Two Bodies,* 47, 49.
18. Quoted in Kantorowicz, *King's Two Bodies,* 48.
19. Derrida, *Beast/Bête,* 1:285/382. The critical passage of the French reads: "Je dis bien le corps vivant d'un roi mais aussi le corps mortel d'un roi qui dispose d'un autre corps, de ce double corps (terrestre, mortel et céleste, sublime, éternel, celui de la fonction ou de la majesté, donc de la souveraineté royale supposée survivre éternellement au corps mortel—et la souveraineté le suppose toujours, de quelque façon et en général, *a priori,* dirai-je, car c'est dans la structure du concept de souveraineté, [qui suppose toujours, donc,] ce double corps dont Kantorowicz, dans *The King's Two Bodies* (1957), aura analysé la structure."
20. Derrida goes on to explain that "Fiction is the *dynamis* and the dynasty of the dynast, but also the radiant *energeia* and *enargeia* of his actions, his powers, his potency. . . . The illusion or fiction to which the reading or watching subject is subjected are the very exercise and effect of sovereign power to which they are subject as subjects of the king. Sovereignty is this narrative fiction or this effect of representation. Sovereignty draws all of its power, all of its potency, i.e. its all-powerful nature, from this simulacrum-effect, this fiction- or representation-effect that is inherent and congenital to it, as it were co-originary." Derrida, *Beast,* 1:289.
21. For another reading of this scene, as putting the "king's ipseity on display," see Dawne McCance, *The Reproduction of "Life Death"* (New York: Fordham University Press, 2019), 97–124. Specifically, McCance pursues the implications of Derrida's generalization of Kantorowicz's "double body" for the sovereign subject of ethics, for whom this duality goes hand in hand with a "thinking of 'life worth more than life' " (112).
22. See Derrida, *Life Death.*
23. Derrida, *Speech and Phenomena,* 54. References to the French are taken from Jacques Derrida, *La voix et le phénomène* (Paris: Presses universitaires de France, 1967), 60–61. The French reads: "Si la possibilité de ma disparition en général doit être d'une certaine manière vécue pour qu'un rapport à la présence en général puisse s'instituer, on ne peut plus dire que l'expérience de la possibilité de ma disparition absolue (de ma

mort) vient m'affecter, survient à un *je suis* et modifie un sujet. Le *je suis* n'étant vécu que comme un *je suis présent,* il suppose en lui-même le rapport à la présence en général, à l'être comme présence. L'apparaître du *je* à lui-même dans le *je suis* est donc originairement rapport à sa propre disparition possible. *Je suis* veut donc dire originairement *je suis mortel. Je suis immortel* est une proposition impossible."

24. Derrida, *Speech and Phenomena/La voix et le phénomène,* 54/61. The French reads: "Pour se servir des distinctions de la 'grammaire pure logique' et de *Logique formelle et logique transcendantale,* il faut spécifier ainsi cette impossibilité: cette proposition a certes un sens, elle constitue un discours intelligible, elle n'est pas *sinnlos.* Mais à l'intérieur de cette intelligibilité, et pour la raison que nous venons d'indiquer, cette proposition est 'absurde' (de l'absurdité de contradiction—*Widersinnigkeit*) et *a fortiori* 'fausse.' Mais comme l'idée classique de vérité, qui guide ces distinctions, est elle-même issue d'un tel dérobement du rapport à la mort, cette 'fausseté' est la vérité même de la vérité. C'est donc à travers d'autres, de tout autres 'catégories' (si l'on peut encore appeler ainsi de telles pensées) qu'il faudrait interpréter ces mouvements."
25. Derrida, *Speech and Phenomena/La voix et le phénomène,* 54–55/61. The French reads: "On peut donc aller plus loin: en tant que langage, 'Je suis celui qui suis' est l'aveu d'un mortel. Le mouvement qui conduit du *Je suis* à la détermination de mon être comme *res cogitans* (donc comme immortalité) est le mouvement par lequel l'origine de la présence et de l'idéalité se dérobe dans la présence et l'idéalité qu'elle rend possible."
26. On the relation between autoaffection and life, see Derrida, *Of Grammatology,* 165, as well as my essay: Adam R. Rosenthal, "Artificial Life, Feeling Machines, and the Text of Deconstruction," *Derrida Today* 16, no. 2 (2023): 129–42.
27. Derrida, *Ear of the Other,* 114.
28. Derrida, *Ear of the Other,* 114–16.
29. Jacques Derrida, "The Almost Nothing of the Unpresentable," in *Points . . . Interviews, 1974–1995,* ed. Elisabeth Weber, trans. Peggy Kamuf (Stanford, Calif.: Stanford University Press, 1992), 78–88, 83.
30. Cf. Jacques Derrida, *The Animal That Therefore I Am,* ed. Marie-Louise Mallet, trans. David Wills (New York: Fordham University Press, 2008), 47.
31. Derrida, *Beast/Bête,* 1:42/72.
32. Derrida, *Beast,* 1:76.
33. See especially Jacques Derrida, "Unconditionality or Sovereignty: The University at the Frontiers of Europe," trans. Peggy Kamuf, *Oxford*

Literary Review 31, no. 2 (2009): 115–31, 123; Derrida, "University without Condition"; and Derrida, *Rogues,* xiii–xiv.

34. Derrida, *Beast,* 1:269.
35. Michael Naas, "The Mother, of All the Phantasms," in Derrida and Joyce: Text and Contexts, eds. Andrew J. Mitchell and Sam Slote (Albany, NY: State University of New York Press, 2013), 163–82, 166. On this, see also Pleshette DeArmitt, *The Right to Narcissism: A Case for an Im-possible Self-Love* (New York: Fordham University Press, 2013), 144–45.
36. Naas, "Mother, of All the Phantasms," 166.
37. This has to do with the "instituting" character of the nonpresence that conditions the phantasm, which we have already explored. For two similar examples, see Derrida, "Night Watch"; Jacques Derrida, *Monolingualism of the Other,* trans. Patrick Mensah (Stanford, Calif.: Stanford University Press, 1998). In the latter, Derrida writes: "This abiding 'alienation' appears, like 'lack,' to be constitutive. But it is neither a lack nor an alienation; it lacks nothing that precedes or follows it, it alienates no *ipseity,* no property, and no self that has ever been able to represent its watchful eye. Although this injunction issues a summons, lastingly, nothing else 'is there' ever to watch over its past or future. This structure of alienation without alienation, this inalienable alienation, is not only the origin of our responsibility, it also structures the peculiarity and property of language. It institutes the *phenomenon* of hearing-oneself-speak in order to mean-to-say. But here, we must say the *phenomenon* as *phantasm*" (25).
38. Jacques Derrida, "History of the Lie: Prolegomena," in *Without Alibi,* ed. and trans. Peggy Kamuf (Stanford, Calif.: Stanford University Press, 2002), 28–70, 28.
39. DeArmitt, *Right to Narcissism,* 144–45.
40. Naas, "Mother, of All the Phantasms," 166.
41. Naas, "Mother, of All the Phantasms," 166.
42. Naas, "Mother, of All the Phantasms," 167.
43. Naas, "Mother, of All the Phantasms," 167.
44. See, variously, Derrida, *Monolingualism;* and Jacques Derrida and Anne Dufourmantelle, *Of Hospitality,* trans. Rachel Bowlby (Stanford, Calif.: Stanford University Press, 2000).
45. Derrida, *Monolingualism,* 46.
46. Derrida, *Of Grammatology/De la grammatologie,* 142–43/205–6.
47. On this problem, see especially Geoffrey Bennington, "Handshake," in *Not Half No End: Militantly Melancholic Essays in Memory of Jacques Derrida* (Edinburgh: Edinburgh University Press, 2011), 65–85.
48. Derrida, *Of Grammatology,* 89.

49. Derrida, *Of Grammatology,* 89.
50. Derrida, *Of Grammatology,* 89; see also 69.
51. Derrida, *Of Grammatology/De la grammatologie,* 49/71–72. Compare also to Derrida, *Speech and Phenomena,* 104; Derrida and Dufourmantelle, *Of Hospitality,* 89–93; and Jacques Derrida, *Paper Machine,* trans. Rachel Bowlby (Stanford, Calif.: Stanford University Press, 2005), 63.
52. Jacques Derrida, *Writing and Difference,* trans. Alan Bass (Chicago: University of Chicago Press, 2017), 375.

Part II

1. Derrida, *Beast/Bête,* 1:43/73. The original French reads: "La souveraineté, âme et donc vie de l'État, respiration artificielle de l'État, n'est posée, instituée, promise, contractée, *artificiellement, comme immortelle* que parce qu'elle est *naturellement mortelle.*"
2. Derrida, *Beast/Bête,* 1:43/73. The original French reads: "C'est la technique prothétique et artificielle qui l'immortalise ou lui assure en tout cas une survie indéfinie."
3. In another use of this phrase, Derrida refers to "the indefinite survival of mystification." Derrida, "History of the Lie," 69. See also Jacques Derrida, "Hostipitality," in *Acts of Religion,* ed. Gil Anidjar (New York: Routledge, 2002), 413.

3. Substance Immortality

1. All translations from Descartes's French are mine unless otherwise noted. René Descartes, *Méditations métaphysiques* (Paris: Garnier-Flammarion, 1979), 107.
2. Descartes, *Méditations métaphysiques,* 117.
3. Descartes, *Méditations métaphysiques,* 125.
4. Descartes, *Méditations métaphysiques,* 217–18.
5. Descartes, *Méditations métaphysiques,* 179. See in particular the following passage: "Nature also teaches me through these feelings of pain, hunger, thirst, etc., that I am not only lodged in my body, like a pilot in his ship, but, beyond this, that I am conjoined to it very tightly and so thoroughly confounded and mixed with it that I am composed as a single whole with it" (179).
6. Descartes, *Méditations métaphysiques,* 62–63.
7. Descartes, *Méditations métaphysiques,* 62. The original French reads: "Nous ne concevons aucun corps que comme divisible, au lieu que

l'esprit, ou l'âme de l'homme, ne se peut concevoir que comme indivisible."

8. Descartes, *Méditations métaphysiques,* 252. See also the summary that Descartes gives of his six meditations at the beginning of the volume: "First of all, in order to know that generally every substance, that is, those things that cannot exist without being created by God, are incorruptible by nature, and can never cease to be, unless they are reduced to nothing by this same God that wishes to deny his ordinary support" (62).
9. Derrida, *Beast,* 1:76. See also these remarks, where it is not only sovereignty that is indivisible: " 'sovereignty' (which is always in principle indivisible)"—but also and exemplarily sovereign decision, "in its very institution, and in the instant proper to it, the act of sovereignty must and can, by force, put an end in a single, indivisible stroke to the endless discussion." Derrida, *Rogues,* xiv, 10. For Nancy, the indivisibility that marks the thinking of God, soul, and sovereignty is also at work within the conception of the individual—now defined by absoluteness and immanence—and must be opposed to what he calls singularity, which comes only from the ecstasy of community. On this, see Jean-Luc Nancy, *The Inoperative Community,* ed. and trans. Peter Conor (Minneapolis: University of Minnesota Press, 1991), 6–7.
10. Descartes, *Méditations métaphysiques,* 156–57. In Latin, this passage reads: "Esse necessarium ut & ante ab aeterno extiterit, & in aeternum sit mansurus."
11. Boethius, *Theological Tractates and The Consolation of Philosophy,* trans. H. F. Steward, E. K. Rand, and S. J. Tester (Cambridge, Mass.: Harvard University Press, 1973), 423.
12. Boethius, *Consolation,* 423–25.
13. St. Augustine, *Confessions,* trans. Henry Chadwick (Oxford: Oxford University Press, 1991), 230.
14. For a dissenting view, see, e.g., Geoffrey Gorham, "Descartes on God's Relation to Time," *Religious Studies* 44, no. 4 (2008): 413–31.
15. For this position, see Hannah Arendt, *The Human Condition* (Chicago: University of Chicago Press, 1958), 7–21.

4. Ontological Immortality

1. Heidegger puts this idea succinctly: "The problem is not one of *how* the '*derived*' *[abgeleitete]* infinite time, in which, the ready-to-hand arises and passes away, becomes *primordial* finite temporality; the problem is rather that of how *in*authentic temporality arises out of finite authentic tem-

porality, and how inauthentic temporality, *as in*authentic, temporalizes an *in*-finite time out of the finite. Only because primordial time is *finite* can the 'derived' time temporalize itself as *infinite.*" Martin Heidegger, *Being and Time,* trans. John Macquarrie and Edward Robinson (Oxford: Blackwell, 1962), 379. References to the German are from Martin Heidegger, *Sein und Zeit* (Tübingen: Max Niemeyer Verlag, 2006).

2. Throughout the following discussion, "life" has a twofold valence. On the one hand, it refers to what is called biological life; on the other hand, it refers to the phenomenological sense of world, which ought to inhere in the former. I will return to the much-debated nature of this relation in Part III.
3. See Hägglund, *Radical Atheism,* 2, 49, 111, 129–31. As Jonas sees it, the Cartesian, atomistic conception of time is itself one of immunity, which in turn allows for the conception of the selfsame: "In the ontological respect, then, the classical Atomist thinks of the unmediated, fixed, but also empty self-identity that is equivalent with immunity to time: with the fact, that is, that the final residue of all reduction exists in each of its bits 'once and for all,' and that these bits are mutually external once for all. The single stubborn particle is then simply and inertly what it is, immediately identical with itself, without need to maintain that self-identity as an act of its existence. Its enduring is mere remaining, not re-assertion of being; and in remaining it is successively 'identical' thanks to the homogenous continuity of the dimensions—space and time—in which the sequence of 'its' states takes place . . . in principle at least, the space-time continuity of the particle must be verifiable throughout. The *identity* of a particle in its temporal career is the *continuity* of its space-time path which we can plot" (234). Hans Jonas, "Biological Foundations of Individuality," *International Philosophical Quarterly* 8, no. 2 (1968): 231–51.
4. Jacques Derrida, *Given Time,* trans. Peggy Kamuf (Chicago: University of Chicago Press, 1992), 102.
5. Derrida, *Of Grammatology,* 71. Compare to "The immortality and perfection of a living being would consist in its having no relation at all with any outside. That is the case with God. God has no allergies" (Derrida, "Plato's Pharmacy," 101); and to "A voice without differance, a voice without writing, is at once absolutely alive and absolutely dead" (Derrida, *Speech and Phenomena,* 102).
6. Nancy, *Inoperative Community,* 15.
7. That is, insofar as notions of society and communion both retain the pure presence of immanence as their hallmarks: "Now the community of human immanence, man made equal to himself or to God, to nature,

and to his own works, is one such community of death—or of the dead. The fully realized person of individualistic or communistic humanism is the dead person. In other words, death, in such a community, is not the unmasterable excess of finitude, but the infinite fulfillment of an immanent life: it is death itself consigned to immanence; it is in the end that resorption of death that the Christian civilization, as though devouring its own transcendence, has come to minister to itself in the guise of a supreme work." Nancy, *Inoperative Community,* 14.

8. Christopher Fynsk, foreword to Nancy, *Inoperative Community,* xv.
9. Judith Butler, *Precarious Life* (New York: Verso, 2006); Todd May, *A Fragile Life* (Chicago: University of Chicago Press, 2017); and Todd May, *Death* (London: Routledge, 2014).
10. Jean-Luc Nancy, *Noli me tangere,* trans. Sarah Clift, Pascale-Anne Brault, and Michael Naas (New York: Fordham University Press, 2008), 45.
11. Weber and Varela argue (against Heidegger) for an essential relationship between the fragility of embodiment and the emergence of world. For them, the values, worldliness, autonomy, and internal teleology that mark all organic life must be understood to result from the self-production of organisms that necessarily maintain themselves in constant states of precarity. The idea of substance immortality would therefore contradict the biological substrate whereby any form of self-organization emerges. Andreas Weber and Francisco J. Varela, "Life after Kant: Natural Purposes and the Autopoietic Foundations of Biological Individuality," *Phenomenology and Cognitive Sciences* 1 (2002): 97–125.
12. Blanchot and Derrida, *Instant of My Death/Demeure,* 4–5. Heidegger too expresses this thought. If ever Dasein loses its incompleteness and becomes "whole," as the effacement of its potentiality, then it ceases to be: "But as soon as Dasein 'exists' in such a way that absolutely nothing more is still outstanding in it, then it has already for this very reason become 'no-longer-Being-there.' Its Being is annihilated when what is still outstanding in its Being has been liquidated. As long as Dasein *is* as an entity, it has never reached its 'wholeness.' But if it gains such 'wholeness,' this gain becomes the utter loss of Being-in-the-world." Heidegger, *Being and Time/Sein und Zeit,* 280/236.
13. One may, of course, identify with this survival of the name. Yet whether or not one does so, the kind of immortality that adheres to names seems distinct from the kind of vitality that adheres to organic beings.
14. Heidegger, *Being and Time/Sein und Zeit,* 284/240. The original German reads: "Das Sterben muß jedes Dasein jeweilig selbst auf sich nehmen. Der Tod ist, sofern er 'ist,' wesensmäßig je der meine. Und zwar bedeutet

er eine eigentümliche Seinsmöglichkeit, darin es um das Sein des je eigenen Daseins schlechthin geht. Am Sterben zeigt sich, daß der Tod ontologisch durch Jemeinigkeit und Existenz konstituiert wird. Das Sterben ist keine Begebenheit, sondern ein existenzial zu verstehendes Phänomen."

15. Heidegger, *Being and Time/Sein und Zeit,* 289/245, translation modified.
16. Heidegger, *Being and Time/Sein und Zeit,* 291/247. For a helpful account of this structure, see Kas Saghafi, "Dying Alive," *Mosaic* 48, no. 3 (2015): 15–26.
17. As Derrida puts it: "This would perhaps be the place or moment to clarify once more the both subtle and decisive stakes of this power of the 'I.' No doubt it will not simply be a case of the relation to self, nor even of a certain automation, an auto-kinetic spontaneity that no one, not even the most negative of minds vis-à-vis the animal, not even Descartes, disallows in the animal. Let me repeat it, every living creature, and thus every animal to the extent that it is living, has recognized in it this power to move spontaneously, to feel itself and to relate to itself. However problematic it be, that is even the characteristic of what lives, as traditionally conceived in opposition to the inorganic inertia of the purely physico-chemical. No one denies the animal auto-affection or auto-motion, hence the self of that relation to the self. But what is in dispute—and it is here that the functioning and the structure of the 'I' count so much, even where the word *I* is lacking—is the power to make reference to the self in deictic or autodeictic terms, the capability at least virtually to turn a finger toward oneself in order to say 'this is I.' . . . Because it is held to be incapable of this autodeictic or auto-referential self-distancing and deprived of the 'I,' the animal will lack any 'I think,' as well as understanding and reason, response and responsibility. The 'I think' that must accompany every representation is this auto-reference as condition for thinking, as thinking itself; that is precisely what is proper to the human, of which the animal would be deprived." Derrida, *Animal,* 94.
18. On the sense of the "as such" as what separates the world-forming nature of man from the world-poor nature of the animal, see Heidegger, *Fundamental Concepts/Gesamtausgabe II,* 245–49/358–62. There, Heidegger explains that the animal is essentially *benommen* (dazed, stunned, or numb) to the extent that it lacks access to beings as such—that is, as presences *(vorhanden)*. Only were it to have the objects of its environment as present beings—only then could it in turn subject them to *Überlegungen* (thoughts or deliberations) and then make them accessible as

manifest *(offenbar)*. This is, to the contrary, the precise role that language plays for man.

19. See, in particular, Derrida, *Animal,* 89–93.
20. And what is "death"? For Heidegger, death is nonbeing. That is, it is the experience of the nothing that is only possible for a being with access to being. Only for a being with ontological difference—access to the difference between beings and being—is death as nonbeing a possibility, and one precisely in the form of an impossible possibility. Heidegger does not discount that nonhuman animals have a sense of death as a present. That is, he does not discount that they may have a concept of death as a functional state that opposes life. For a functional account of death for nonhuman animals, see Susana Monsó, "How to Tell If Animals Can Understand Death," *Erkenntnis* 87 (2022): 117–36; and Susan Monsó and Antonio J. Osuna-Mascaró, "Death Is Common, So Is Understanding It: The Concept of Death in Other Species," *Synthese* 199 (2021): 2251–75.
21. Thus, while *Sterben* is irreducible either to *Ableben* or to *Verenden*—*Sterben* names the experience of the possibility of death *(Tod)* as a possibility, always only possible insofar as Dasein exists as Dasein—it nevertheless requires reference to some kind of *Ableben* or *Verenden.* It implies, we could say, some possibility of an event of *Ableben* or *Verenden* as a factical eventuality.
22. The world poverty of the animal, according to Heidegger, consists in a fundamental *Benommenheit,* or numbness, that it has with respect to the being of beings. The animal has access to beings—indeed, it lives through an exposure to beings—but, says Heidegger, it cannot approach these beings as such; it cannot form representations of them apart from their immediate appearance. This is the privilege of language, the logos, which allows the being of beings to come to appearance. Man, in contrast, not only has world but is also world making, insofar as his access to beings through the logos allows for valid as well as invalid reflection on them.
23. Heidegger, *Being and Time/Sein und Zeit,* 284/240. Perhaps the animal can learn to "expect death" *(im Erwarten)* as the arrival of a present event, but according to Heidegger, it could not "anticipate death," by which he means, anticipate its possibility *(Vorlaufen in die Möglichkeit)* as a possibility, and one that lurks perpetually, and in fact makes possible its expectation as an actuality (306–7/261–63).
24. Derrida, *Animal,* 129. Derrida, *Beast,* expresses the same thought: "As opposed to the non-death, if not the immortality, of the animal" (2:103); "one can read 'Das Ding' as a great text on death, on the mortality of

Dasein, in opposition to an a-mortality or even an immortality of the beast" (2:120). See also 2:124 and 2:157; and see Jacques Derrida, *Aporias,* trans. Thomas Dutoit (Stanford, Calif.: Stanford University Press, 1993), 35–38, 39–40, 55, 74–76.

25. Martin Heidegger, "The Nature of Language," in *On the Way to Language,* trans. Peter D. Hertz (San Francisco, Calif.: Harper & Row, 1982), 107, translation modified. References to the German are taken from Martin Heidegger, *Gesamtausgabe I. Abteilung: Veröffentlichte Schriften, 1910–1976. Band 12. Unterwegs zur Sprache,* (Frankfurt am Main: Vittorio Klostermann, 1985), 203.
26. Heidegger, *Fundamental Concepts/Gesamtausgabe II,* 2:238/346.
27. Jean-Jacques Rousseau, *The Social Contract and the First and Second Discourses,* ed. Susan Dunn (New Haven, Conn.: Yale University Press, 2002), 97. Rousseau is less explicit as to the animal's immortality per se than he is of its ignorance of death. The animal, like the savage, has only needs but no passions. Thus, while self-preservation is an issue for both, it is only for man, with his factitious *amour-propre,* that the idea of immortality emerges.
28. John Keats, *Lamia, Isabella, the Eve of St. Agnes, and Other Poems* (London: Taylor and Hessey, 1820), lines 15–30, 61–62.
29. The presumption of the *Benommenheit* of the animal, understood to be a consequence of its ontological immortality, can also lead to the determination of a form of species immortality. Species immortality, which depends on an effacement of the difference between individuals, is in turn closely linked to the clonal forms of immortality that we will pursue in the following chapters.
30. Heidegger, *Being and Time/Sein und Zeit,* 310/265.
31. Heidegger, *Being and Time/Sein und Zeit,* 280/236.

5. Biological Immortality

1. "In our terminology the ending of anything that is alive, is denoted as 'perishing.'" Heidegger, *Being and Time/Sein und Zeit,* 284/240. In German: "Das Enden eines Lebendigen fassen wir terminologisch als Verenden." "In so far as this is the case, Dasein too can end without authentically dying, though on the other hand, *qua* Dasein, it does not simply perish. We designate this intermediate phenomenon as its '*demise*'" (291/247). "Sofern auch das Dasein seinen physiologischen, lebensmäßigen Tod 'hat,' jedoch nicht ontisch isoliert, sondern mitbestimmt durch seine ursprüngliche Seinsart, das Dasein aber auch enden kann,

ohne daß es eigentlich stirbt, andererseits qua Dasein nicht einfach verendet, bezeichnen wir dieses Zwischenphänomen als *Ableben*."

2. William Wordsworth, "Essay upon Epitaphs," in *The Prose Works of William Wordsworth*, vol. 2, ed. W. J. B. Owen and Jane Worthington Smyser (Oxford: Clarendon, 1974), 57.
3. The fields of biological life and mortal life of course need not coincide.
4. Jean-Baptiste de Monet de Lamarck, *Philosophie zoologique: Tome premier*, ed. Charles Martins (Paris: Libraire F. Savy, 1873), 105–6. The translation is my own.
5. And thus today we find the idea of a "genetically predetermined death," as determined by concepts of molecular biology. On this and its evolutionary link to the development of sex, see, e.g., Jacob, *Logic of Life*, 309–10; and Nick Lane, *The Vital Question* (New York: Norton, 2015), 14.
6. Plato, *Timaeus*, 242–43. On this, see also Naas, *Plato*, 240n10.
7. Heidegger, *Being and Time*, 301.
8. The discovery in 1961 of the Hayflick limit was supposed to demonstrate the internal upper limit on the number of cell divisions possible for any human cell before apoptosis (programmed cell death). Many life-extension advocates and gerontologists, such as David Sinclair and Aubrey de Grey, seek to find workarounds to this limit. See L. Hayflick and P. S. Moorhead, "The Serial Cultivation of Human Diploid Cell Strains," *Experimental Cell Research* 25, no. 3 (1961): 585–621.
9. For an overview of some of the problems confronting notions of self and individual from a biological perspective, see Michael Levin, "Life, Death, and Self: Fundamental Questions of Primitive Cognition Viewed through the Lens of Body Plasticity and Synthetic Organisms," *Biochemical and Biophysical Research Communications* 563, no. 30 (2021): 114–33; Chris Fields and Michael Levin, "Are Planaria Individuals? What Regenerative Biology is Telling Us about the Nature of Multicellularity," *Evolutionary Biology* 45 (2018): 237–47; and Samuel L. Díaz-Muñoz, Amy M. Boddy, Gautam Dantas, Christopher M. Waters, and Judith L. Bronstein, "Contextual Organismality: Beyond Pattern to Process in the Emergence of Organisms," *Evolution* 70, no. 12 (2016): 2669–77. In brief, organismal individuality is not only a problem from an evolutionary perspective but also from an ontogenetic, developmental one; for example, some slime molds are able to pass through multiple single- and multicellular configurations during a "single" life cycle. I will return to these problems and their implications for human life and individuality in Parts III and IV.
10. For Godwin's comments on immortality—or, more precisely, the "prolongation of human life"—see William Godwin, *Enquiry Concern-*

ing Political Justice and Its Influence on Modern Morals and Happiness (London: G. G. and J. Robinson, 1798), 2:519–29. For Condorcet's thoughts on increasing the "mean duration of human life," see Jean-Antoine-Nicolas de Condorcet, *Outlines of an Historical View of the Progress of the Human Mind* (London: J. Johnson, 1795), 366–72. For Malthus's response to both, see Thomas Robert Malthus, *An Essay on the Principle of Population,* ed. Geoffrey Gilbert (Oxford: Oxford University Press, 1993), 67–73, 99–102. See also McLane's excellent discussion of the changing senses of life and death in the eighteenth and nineteenth centuries: Maureen McLane, "Dead Poets and Other Romantic Populations: Immortality and Its Discontents," in *Romanticism and the Human Sciences* (Cambridge: Cambridge University Press, 2000), 159–215.

11. McLane, "Dead Poets," 168–69.
12. Godwin differentiates between the "absolute immortality of man"—or something similar to what I call substance immortality above—and this indefinite prolongation of life on earth as a potential made possible through perfectibility. Godwin, *Enquiry Concerning Political Justice,* 527. Malthus, by contrast, moves back and forth between the language of prolongation and that of immortality, precisely to emphasize their underlying connection.
13. McLane, "Dead Poets," 172.
14. Malthus, *Essay on the Principle of Population,* 101–2.
15. William Hazlitt, *A Reply to the Essay on Population* (London: Longman, Hurst, Rees, and Orme, 1807), 248.
16. This term is Ian Wilmut and Keith Campbell's—the scientists responsible for cloning Dolly the sheep. See Ian Wilmut, Keith Campbell, and Colin Tudge, *The Second Creation* (London: Headline, 2000), 17.
17. Sarah Franklin, *Dolly Mixtures* (Durham, N.C.: Duke University Press, 2007), 33.

6. Mortality Revisited

1. Derrida, *Death Penalty,* 253. References to the French are taken from Jacques Derrida, *Séminaire: La peine de mort, Volume 1 (1999–2000)* (Paris: Galilée, 2012), 344.
2. See Heidegger, *Being and Time,* 306, 309.
3. Derrida, *Death Penalty,* 258. On the question of time and the death penalty, see also David Wills, *Killing Times: The Temporal Technology of the Death Penalty* (New York: Fordham University Press, 2019).
4. Derrida, *Death Penalty,* 256–57/347–48.

5. It is thus no longer simply a question of the *when* of death, which uncertain "when" had made its time of arrival indefinite, but now a question of the *if* of death, which makes indefinite the fact of its arrival as such.
6. As Derrida notes, however, in the second half of the Death Penalty seminars, the question of the condemnation to death (or death penalty) is notably absent from Heidegger's writings.
7. Heidegger, *Being and Time/Sein und Zeit,* 302/258. See also the following passages: "In anticipating the indefinite certainty of death, Dasein opens itself to a constant threat arising out of its own 'there.' In this very threat, being-toward-the-end must maintain itself. So little can it tone this down that it must rather cultivate the indefiniteness of the certainty" (310/265). "In the 'sometime, but not right away,' everydayness concedes something like a *certainty* of death. Nobody doubts that one dies. On the other hand, this 'not doubting' need not imply *that* kind of Being-certain which corresponds to the way death—in the sense of the distinctive possibility characterized above—enters into Dasein. Everydayness confines itself to conceding the 'certainty' of death in this ambiguous manner just in order to weaken that certainty by covering up dying still more and to alleviate its own thrownness into death" (299–300/255–56).
8. Heidegger, *Being and Time/Sein und Zeit,* 303/259.
9. Heidegger, *Being and Time/Sein und Zeit,* 252/296, 253/297.
10. Yuval Noah, *Sapiens* (New York: Harper, 2015), 385, 413. Derrida, by contrast, refers to the animal as "amortal," in reference to its ontological immortality.

7. Undecidable Desire for Survival

1. Yuval Noah Harari, "Harari: On Homo Deus, Immortality, Dataism and Health, the 'Infinite Market,'" interview conducted by Zulfikar Abbany, *DW,* February 9, 2016, https://www.dw.com/en/harari-on-homo-deus-immortality-dataism-and-health-the-infinite-market/a-19523293.
2. Sixty years ago, Arendt had already concluded that the key to the difference in approaches lies in a historical shift relating to the modern recognition of the impermanence of reality. This shift would have forced us moderns to modify our aspirations but perhaps not actually to amend their underlying motives. See Arendt, *Human Condition,* 21; Hannah Arendt, "The Concept of History: Ancient and Modern," in *Between Past and Future* (New York: Meridian, 1961), 41–90, 72–73; and Hannah Arendt, "History and Immortality," in *Partisan Review* 24, no. 1 (1957):

11–35. For a careful examination of the problem of immortality in Arendt's work, as always tied to the immortality of the world (and not the individual), see Peg Birmingham, "Worldly Immortality in an Age of Superfluity," *Arendt Studies* 2 (2018): 25–36.

3. Harari, *Sapiens,* 403.
4. See Nick Bostrom, "Ethical Issues in Advanced Artificial Intelligence," in *Science Fiction and Philosophy,* ed. Susan Schneider (Oxford: Wiley-Blackwell, 2009), 277–84, 282. Bostrom later repeats this prediction, along with the obligation to pursue it, elsewhere: "If the machine intelligence revolution goes well, the resulting superintelligence could almost certainly devise means to indefinitely prolong the lives of the then still-existing humans, not only keeping them alive but restoring them to health and youthful vigor, and enhancing their capacities well beyond what we currently think of as the human range; or helping them shuffle off their mortal coils altogether by uploading their minds to a digital substrate." Nick Bostrom, *Superintelligence,* 245–46. As he explains, although "the human desire to acquire new capacities is as ancient as our species itself," the roots of transhumanism are to be found in rational humanism and Enlightenment thought. Nick Bostrom, "A History of Transhuman Thought," *Journal of Evolution and Technology* 14, no. 1 (2005): 1–25, 1.
5. On this, see Nick Bostrom, "Letter from Utopia," in *Studies in Ethics, Law, and Technology* 2, no. 1 (2008): 1–7. Incidentally, Godwin had already proposed such a possibility of limitation: "The rational anticipations of human improvement are unlimited, not eternal. The very globe that we inhabit, and the solar system, may, for anything that we know, be subject to decay." Godwin, *Enquiry Concerning Political Justice,* 518.
6. See Bernard Williams, *Problems of the Self* (Cambridge: Cambridge University Press, 1973); Nussbaum, *Fragility of Goodness;* May, *Death;* and May, *Fragile Life.*
7. Hägglund, *Radical Atheism,* 131.
8. "The point to underline here is that Socrates does not say that the man wants to transcend his condition of mortal health. On the contrary, he wants to *go on* being what he is. And since he is mortal, he wants to *live on* as mortal. As we will see, this desire for survival is incompatible with the desire for immortality, since it wants to hold on to a life that is essentially mortal and inherently divided by time." Hägglund, *Dying for Time,* 5. For critics of Hägglund's position, see especially Johnston, "Life Terminable and Interminable"; and Kas Saghafi, *The World after the End of the World* (Albany, N.Y.: SUNY Press, 2020).
9. Hägglund, *Radical Atheism,* 1.

10. Hägglund, *This Life,* 6.
11. Bostrom, "Letter from Utopia," 3–4.
12. Perhaps it would be better to say: if Descartes's aim of immortality is suicidal, then such a suicide appears to be more complex than mere annihilation. It is a suicide that lasts.
13. Derrida, *Death Penalty,* 279.
14. For a slightly different account of the resilience of religion, as the consequence of faith that springs from the fabulous origin of every performative act, see Jacques Derrida, "Faith and Knowledge," in *Acts of Religion,* 56–57. Further: "Religion and reason develop in tandem, drawing from this common resource: the testimonial pledge of every performative, committing it to respond as much *before* the other as *for* the high-performance performativity of technoscience. The same unique source divides itself mechanically, automatically, and sets itself reactively in opposition to itself: whence the two sources in one. This reactivity is a process of *sacrificial indemnification,* it strives to restore the unscathed *(heilig)* that it itself threatens. And it is also the possibility of the two, of $n + 1$, the same possibility as that of the *testimonial deux ex machina*" (66). See also Derrida, *Death Penalty,* 254–59.

Part III

1. For an examination of the relationship between Derridean deconstruction and the life sciences, see Francesco Vitale, *Biodeconstruction* (Albany, N.Y.: SUNY Press, 2018). On the problem of life in general in deconstruction, see Mauro Senatore, *Germs of Death* (Albany, N.Y.: SUNY Press, 2018); and Robert Trumbull, *From Life to Survival* (New York: Fordham University Press, 2022). In what follows, I depart from their work insofar as my question is not about the problem of life encountered in Derrida's thinking but is rather this: how do the life sciences themselves reproduce the problems that we have already encountered in Plato and Derrida?
2. On the life sciences' presupposition of an "individual living unity," see Leonardo Bich, "Complex Emergence and the Living Organization," *Synthese* 185 (2012): 215–32, 217.
3. As Jonas writes, what materialism achieves is the reduction of all life to death. Far from resolving the issue, however, this simplification merely leads to the identification of the living with its observable processes and death with their absence. See Jonas, "Life, Death, and the Body in the Theory of Being," in *Phenomenon of Life,* 7–25.

4. On the supreme difficulty of thinking "life" nonoppositionally, see Derrida, *Life Death.*
5. In this way, we could say that the organism itself becomes "postvital," to adopt Doyle's term. That is, having become fully transparent by way of its reduction to molecular structures and physiological functions, it is transformed into a pure presence with "nothing behind or beyond it" (13). See Richard Doyle, *On Beyond Living: Rhetorical Transformations of the Life Sciences* (Stanford, Calif.: Stanford University Press, 1997).
6. To admit as much is not the same as accepting that there is a hard-and-fast divide between some living as such and some nonliving as such. I shall try to approach this problem gradually in what follows.
7. The range of scholarship on this problem is immense and spans highly disparate disciplines. On the biological foundations of identity in minded as well as unminded life-forms, see in particular Antonio Damasio, *The Strange Order of Things* (New York: Pantheon, 2018). On life and basal cognition, see Levin, "Life, Death, and Self," 4. For a controversial account of mind that links consciousness with cellularity, see Arthur Reber, *The First Minds* (Oxford: Oxford University Press, 2018).
8. For various notable efforts to link the phenomenon of life with organization, information, and complexity, see the following: Sara Walker, Paul Davies, and George Ellis, eds., *From Matter to Life* (Cambridge: Cambridge University Press, 2017); Alvaro Moreno and Matteo Mossio, *Biological Autonomy* (Dordrecht: Springer, 2015); and Stuart Kauffman, *A World beyond Physics* (Oxford: Oxford University Press, 2019).

8. Genetic Immortality

1. For various critiques of this position, see Hayles, *How We Became Posthuman;* Jean-François Lyotard, "Can Thought Go On without a Body?," trans. Brune Boone and Lee Hildreth, *Discourse* 11, no. 1 (1988–89): 74–87; Evan Thompson, *Mind in Life* (Cambridge, Mass.: Harvard University Press, 2007), 166–218; and Susan Oyama, *The Ontogeny of Information* (Durham, N.C.: Duke University Press, 2000).
2. On life as a planetary phenomenon, see, e.g., Tobias Owen, "Life as a Planetary Phenomenon," in *Origins of Life and Evolution of the Biosphere* 15 (1985): 221–34; James Lovelock and Lynn Margulis, "Atmospheric Homeostasis by and for the Biosphere: The Gaia Hypothesis," *Tellus* 26, no. 1–2 (1974): 2–10; and Adam Frank, David Grinspoon, and Sara Walker, "Intelligence as a Planetary Scale Process," *International Journal of Astrobiology* 21 (2022): 47–61.

3. See Carol Cleland, *The Quest for a Universal Theory of Life* (Cambridge: Cambridge University Press, 2019). For an excellent discussion of the deconstructive implications of astrobiological discourse, see Armando Mastrogiovanni, "Exoheliotrope: Metaphor in the Texts of Astrobiology and Deconstruction," *Oxford Literary Review* 45, no. 2 (2023): 208–28.
4. NASA, "About Life Detection," n.d., https://astrobiology.nasa.gov/research/life-detection/about/. For a representative discussion of the limits of this definition, see Stuart Bartlett and Michael Wong, "Defining Lyfe in the Universe: From Three Privileged Functions to Four Pillars," *Life (Basel)* 10, no. 4 (2020): 42.
5. At any rate, among the known beings that have survived to this point, they are always deeply entangled. It is not just entirely conceivable but entirely probable that there have been many metabolically successful organisms that have lived during the last four billion years on earth but that did so without the genetic machinery required to propagate themselves. The subset of organisms we have available to us is missing these shadow generations, so our conception of life is biased by this blindness. The teleological orientation presupposed to be an essential feature of life—to survive and reproduce—is at least partially an observation made possible by the limited sample of organisms that remain, and that remain because they fulfill this requirement.
6. On this distinction, see Sara Walker and Michael Lachmann, "Life ≠ Alive," *Aeon*, June 24, 2019, https://aeon.co/essays/what-can-schrodingers-cat-say-about-3d-printers-on-mars. In short, while all artifacts produced by organisms are, by virtue of their origin, a part of life, they are not necessarily alive. Aliveness minimally entails the further fact of acting as a constructor. The way that metabolically active organisms construct has everything to do with how they live (i.e., are alive). Likewise, for genes, their mode of participation in construction dictates their difference from organismal totalities on the one hand, and their difference from instances of nonlife on the other. Ultimately, there may be no hard-and-fast limit between mere life and aliveness, as we shall see, but this does not detract from its utility as a heuristic.
7. August Weismann, *Essays upon Heredity and Kindred Biological Problems*, ed. Edward B. Poulton, Selmar Schönland, and Arthur E. Shipley (Oxford: Clarendon, 1889), 122.
8. Weismann, *Essays upon Heredity*, 180.
9. Weismann, *Essays upon Heredity*, 168.
10. For a critical appraisal of the importance of Weismann's work, as well as an explanation of its flaws for understanding embryology, see Leo Buss,

The Evolution of Individuality (Princeton, N.J.: Princeton University Press, 1988). The glaring issue with Weismann's conception is that, because he is constrained to locating heredity at the level of the germ cell, he could not account for the highly variable and evolutionarily differentiated embryological processes found in nature. By locating heredity at the gene level, subsequent developmental biologists were able to circumvent this issue.

11. Richard Dawkins, "Introduction to the 30th Anniversary Edition," in *The Selfish Gene* (Oxford: Oxford University Press, 2016), x.
12. Dawkins, *Selfish Gene,* 42–43. In addition: "Genes, like diamonds, are forever, but not quite in the same way as diamonds. It is an individual diamond crystal that lasts, as an unaltered pattern of atoms. DNA molecules don't have that kind of permanence. The life of any one physical DNA molecule is quite short—perhaps a matter of months, certainly not more than one lifetime. But a DNA molecule could theoretically live on in the form of copies of itself for a hundred million years. . . . What I am doing is emphasizing the potential near-immortality of a gene, in the form of copies, as its defining property" (44).
13. See David Haig, *From Darwin to Derrida* (Cambridge, Mass.: MIT Press, 2020), 23.
14. Dawkins, *Selfish Gene,* 26.
15. Daniel Dennett, *Consciousness Explained* (Boston, Mass.: Little, Brown, 1992), 205.
16. For the immortality of DNA, and by extension the organism considered as information, see Richard Dawkins, *River Out of Eden* (London: Weldenfelt & Nicolson, 1995), 4; Richard Dawkins, *The Blind Watchmaker* (New York: Norton, 1986), 173; and Dennett, *Consciousness Explained,* 367–68, 430.

9. Biological Immortality I

1. On this distinction, see Derek Parfit, *Reasons and Persons* (Oxford: Oxford University Press, 1984), 201–2.
2. For Deutsch, all knowledge constitutes "abstract constructors." See David Deutsch, "Constructor Theory," *Synthese* 190, no. 18 (2013): 4331–59.
3. See Kauffman, *World beyond Physics,* 6–11. For a helpful account of the importance of this Kantian notion of organization for modern philosophy and biology, see Yuk Hui, *Recursivity and Contingency* (Lanham, Md.: Rowman & Littlefield, 2019).
4. Haig, *From Darwin to Derrida,* xxv–xxvi.

5. This is the point at which genetics departs from Derridean iterability. Iterability states the minimal conditions of possibility for signification through the potential, inherent in any mark, to be reproduced. In order for any mark to appear as a mark, it must be iterable—that is, caught up in a chain of signifiers from which it could be displaced. By contrast, where the meme takes off for Dawkins and Haig, or where the Kantian whole is achieved for Kauffman, is the point at which a given trace positively affects the possibility of its future (and past) reproductions. To read a trace through (or as) such a teleological orientation, toward its own reproduction, may be a (historical) possibility of iterability, but it precisely misses the essential contingency and nonidentity at the heart of every mark.
6. To be clear, as Parfit explains, "numerical identity" need not imply "exact similarity." Parfit, *Reasons and Persons*, 241. I use it here and elsewhere to describe some material or physiological continuity, in distinction to "qualitative identity," which implies none. As we shall see, above all in Part IV, because every determination of numerical identity implies a kind of qualitative evaluation, in certain cases, the distinction between numerical and qualitative forms of identity breaks down.
7. Jacob, *Logic of Life*, 297. In fact, the French is stronger: "Les bactéries s'y reproduisent éternellement."
8. Jacob, *Logic of Life*, 297.
9. Unlike eukaryotes, prokaryotic bacteria appear not to be subject to a Hayflick limit on divisions. There are many variations on this process, however, including both binary and multiple fission (in which a plurality of daughter cells are produced at once). On some of the diverse forms of bacterial reproduction, see John Waterbury and Roger Stanier, "Patterns of Growth and Development in Pleurocapsalean Cyanobacteria," *Microbiology Reviews* 42, no. 1 (1978): 2–44.
10. Baudrillard, *Vitalist Illusion*, 6–7.
11. Georges Bataille, *The Accursed Share, Volume 1: Consumption* (New York: Zone, 1988), 32.
12. Michael R. Rose, *The Long Tomorrow: How Advances in Evolutionary Biology Can Help Us Postpone Aging* (Oxford: Oxford University Press, 2005), 101–2.
13. On this phenomenon, see Wiep Klaas Smits, Oscar P. Kuipers, and Jan-Willem Veening, "Phenotypic Variation in Bacteria: The Role of Feedback Regulation," *Nature Reviews Microbiology* 4 (2006): 259–71. For an examination of the phenomenon more broadly in developmental biology, see Scott F. Gilbert, "The Genome in Its Ecological Context:

Philosophical Perspectives on Interspecies Epigenesis," *Annals of the New York Academy of Sciences* 981, no. 1 (2002): 202–18. As Gilbert explains: "Developmental plasticity (sometimes called phenotypic plasticity) is the notion that the genome enables the organism to produce a range of phenotypes. There is not a single phenotype produced by a particular genotype. The structural phenotype instructed by the environmental stimulation is referred to as a morph. When developmental plasticity manifests itself as a continuous spectrum of phenotypes expressed by a single genotype across a range of environmental conditions, this spectrum is called the norm of reaction (or reaction norm)" (208).

14. I here adopt the terminology of Maturana and Varela, who differentiate between an organization that persists over time and a structure that serves the organization but that may itself change. Humberto Maturana and Francisco Varela, *The Tree of Knowledge* (Boston, Mass.: Shambala, 1987).
15. Were one nevertheless to seek an element of material stability amid all this change, it would be the individual strand of DNA. As was discovered in 1957 by Matthew Meselson and Franklin Stahl, with each event of DNA replication, there is semiconservative preservation of the genome. In other words, while the two strands of the parental DNA helix are always split among their progeny (that is, between the two new cells after division), each strand is nevertheless preserved integrally together—what Jacob refers to as the " 'old' chain" and the " 'new' one." Jacob, *Logic of Life,* 297. Thus, if one wished to entertain the immortality of bacteria yet evade the paradoxes that result from it, one could perhaps try to do so by reducing "the bacterium" to the single helical strand. This strand constitutes a material gene—in the parlance of Haig—as opposed to the informational gene that is shared between each of the offspring. Because there is only one instance of any given material gene, and because this strand is preserved from generation to generation (it always passes to exactly one offspring within each act of maternal division), it is both immortal and singular. Interesting though it may be, the individual strand of DNA is neither what biologists refer to when they speak of the immortality of bacteria (for it does not encompass a physiologically self-sustaining whole), nor what neo-Darwinists refer to when they speak of the immortality of the gene (for being of the material variety, it is more restrictive than the informational gene, which is equally present in all the bacterial offspring). Thus, it is doubtful that it could solve any of the philosophical issues that follow from bacterial reproduction. Though interesting, it is insufficient, for it can only resolve the aporia of clonal duplication at the

cost of dismissing what our intuition tells us are the essential elements that most need accounting for: informational genes and physiologically self-sustaining entities.

16. Weismann, *Essays upon Heredity,* 25–26.
17. Leonard Hayflick, "The Illusion of Cell Immortality," *British Journal of Cancer* 83, no. 7 (2000): 841–46, 841.
18. See also Souvik Bhattacharyya, "Do Bacteria Age? A Contemporary Controversy in Biology," *Resonance* (2012): 347–64. Evolutionary biology conventionally holds that while eukaryotes age, prokaryotes do not. The reason for this is the separation of germ cells from somatic cells for the latter. As Bhattacharyya puts it, "Aging [is understood to be] a compromise to pass on undamaged genetic material and other cellular components to the next generation. To keep germ cells undamaged, somatic cells make a trade-off with Nature and start aging. So, from this perspective, bacteria do not age" (351–52). However, as he goes on to show, various factors, including asymmetrical cell division, can lead to the natural death of individual bacteria, as was shown to take place for *Caulobacter cresentus* after approximately 130 divisions. On the strategic placement of accrued cellular damage in one (versus both) daughter cell, see Camilla U. Rang, Annie Y. Peng, and Lin Chao, "Temporal Dynamics of Bacterial Aging and Rejuvenation," *Current Biology* 21, no. 21 (2011): 1813–16. This aging apparently leads to a slowing in future divisions, and eventually death.
19. Parfit, *Reasons and Persons,* 256.
20. Hayflick, "Illusion of Cell Immortality," defines it in the following two ways: "a more liberal definition of biological immortality would be the indefinite survival of a life form whose vital life processes function indefinitely" (842); and "in its simplest, though still inaccurate form, biological mortality is usually defined as the death of an organism or the termination of its lineage. Immortality would be defined as the indefinite survival of a single organism or of a replicating population regardless of molecular turnover" (842).
21. It matters little, for the moment, whether this personal identity is defined through psychological continuity, as Parfit develops, or through the continuity of a differentiated, embodied, multicellular organism, as we shall explore in chapter 11.
22. Michael Levin, "Michael Levin on Growth, Form, Information, and the Self," in *Sean Carroll's Mindscape,* podcast, 1:21:33, episode 132, February 2021, https://www.preposterousuniverse.com/podcast/. On planarian immortality, see Levin, "Are Planaria Individuals?"

23. Damasio, *Strange Order of Things,* 40.
24. Sara Walker (@Sara_Imari), post on X (formerly Twitter), January 6, 2022, 11:08 a.m., https://twitter.com/Sara_Imari/status/1479122723087347727.
25. Lynn Margulis, "Kingdom Animalia: The Zoological Malaise from a Microbial Perspective," *American Zoologist* 30, no. 4 (1990): 861–75, 869.
26. Jean Baudrillard, *Simulacra and Simulation,* trans. Sheila Faria Glaser (Ann Arbor: University of Michigan Press, 1994), 96.
27. On the astrobiological tendency to unify all of known life, see Sara Walker and Paul C. W. Davies, "The 'Hard Problem' of Life," in Walker, Davies, and Ellis, *From Matter to Life,* 19–37, 20.
28. Nancy, *Noli me tangere,* 45.

10. Interlude on Death, Division, and the Life of Henrietta Lacks

1. Rebecca Skloot, *The Immortal Life of Henrietta Lacks* (New York: Crown, 2010), 1.
2. For a helpful analysis of the history and biopolitical implications of immortal cell lines, see Margaret Lock, "The Alienation of Body Tissue and the Biopolitics of Immortalized Cell Lines," *Body and Society* 7, no. 2–3 (2001): 63–91.
3. Skloot, *Immortal Life,* 30.
4. Skloot, *Immortal Life,* 41.
5. Skloot, *Immortal Life,* 7, 161, 6, 180, 181, 182, 186.
6. How else to understand the dying wish of George Gey, the biologist responsible for the acquisition and dissemination of HeLa cells? "Before going into the operating room," on having learned that he had pancreatic cancer, "George told his surgeons that he wanted them to take samples of his tumor, just as Dr. Wharton had done with Henrietta's tumor decades earlier. Gey gave his lab staff careful instructions for growing GeGe, a line of cancer cells taken from his pancreas. He hoped that his cells, like Henrietta's, would become immortal." Skloot, *Immortal Life,* 171.

11. Biological Immortality II

1. That is, between any two generations. Great variation can of course arise over extended lengths of time.
2. By "genetically determined" event of death, one need not understand an explicit or even teleological program of senescence or apoptosis. "Genetically determined" may simply signify that the organism's DNA

does not provide it with adequate tools to live past a certain age. In other words, it may simply be that longer vitality has not been selected for in sexed organisms because extensions past a sexual prime do not significantly affect reproduction rates.

3. For an exception to the unicellular/multicellular binary, see Levin, who explains how *Dictyostelium* alternates between "an amoeba phase, and a multicellular phase where the amoebas gather together into a body structure." Levin, "Life, Death, and Self," 8.
4. There is, in sum, no biological immortality tout court. In each case, limits inherent to the nature of the organism impose restrictions on its possible determinations. In distinction to genetic immortality, then, which is regulated by an ideal of immutability, biological immortality is always differential: a matter of degrees, dictated by idiomatic morphologies. As we shall see, one consequence of the differential character of biological immortality is that one can neither determine, absolutely, its presence, nor determine, absolutely, its absence.
5. Interestingly, though it was long believed that this was true for grafts as well, studies have shown that there is horizontal gene transfer between scion and rootstock. The resulting chimeras bear novel genomes. On this, see Jing Wang, Libo Jiang, and Rongling Wu, "Plant Grafting: How Genetic Exchange Promotes Vascular Reconnection," *New Phytologist* 214, no. 1 (2017): 56–65.
6. The basic regenerative capacity of plants is made possible by one of two causes: either the presence of meristems and stem cells throughout their bodies, or the capacity to dedifferentiate specialized cells into totipotent cells. There is thus no hard-and-fast division between soma and germ line—or, at any rate, none that persists for asexual clonal propagation. One consequence of this enhanced regenerative capacity is that "new" plants resulting from sprays preserve whatever genetic mutations happen to be present in the somatic stem cells from which they grow. This means that although there is no meiotic recombination of the genome in sex, a high degree of genetic variance nevertheless arises between donor and cloned plant. On this, see Caifu Jiang et al., "Regenerant *Arabidopsis* Lineages Display a Distinct Genome-wide Spectrum of Mutations Conferring Variant Phenotypes," *Current Biology* 21 (2011): 1385–90.
7. On the problem of the organism and the difficulty of furnishing a definition, see Díaz-Muñoz et al., "Contextual Organismality"; David C. Queller and Joan E. Strassmann, "Beyond Society: The Evolution of Organismality," *Philosophical Transactions of the Royal Society B* 364 (2009): 3143–55; and Levin, "Life, Death, and Self." For a similarly

contextual approach to the individuality of viruses—and the related question of their status as living or nonliving—see Patrick Forterre, "To Be or Not to Be Alive: How Recent Discoveries Challenge the Traditional Definitions of Viruses and Life," *Studies in History and Philosophy of Biology and Biomedical Sciences* 59 (2016): 100–108.

8. This is of course a simplified picture of forms of reproduction. One can also have "sexual" reproduction with only one parent. This form of self-fertilization, or autogamy, is found widely among plants as well as protists. It involves cells of a single organism undergoing meiosis and then genetic recombination with only one parent. For our purposes, it poses no significant difference to sexual reproduction featuring multiple parents.
9. On this, see in particular Howard Thomas, "Do Green Plants Age, and If So, How?," *Topics in Current Genetics* 3 (2004): 145–71. If we allow that there is some degree of autonomous self-regulation that takes place for the holobiont, then there is unity, first, at the level of the whole (composed of the totality of metamers contiguously connected); second, at the level of the individual units (metamers) within that body; and third, at the level of the individual cells composing each of those units. Each of these levels of integration depends on the total environment in which they persist, including the symbionts that make their lives possible. Nevertheless, at each level of organization, there is what we could call cognition, and thus effect of identity.
10. Immanuel Kant, *Critique of the Power of Judgment,* ed. Paul Guyer, trans. Paul Guyer and Eric Matthews (Cambridge: Cambridge University Press, 2000), 243–44. On the history of the problem, see Michael Marder, *Plant-Thinking: A Philosophy of Vegetal Life* (New York: Columbia University Press, 2013).
11. Chris Fields and Michael Levin, "Are Planaria Individuals? What Regenerative Biology Is Telling Us about the Nature of Multicellularity," *Evolutionary Biology* 45 (2018): 237–47.
12. The ontological immortality of Keats's bird, explored in chapter 4, can thus lead to the further conclusion that all members of the species are identical, and on this ground, that events of death and birth are effectively nonevents because there is preservation of the type despite the loss of tokens. Knowledge of clonal versus sexual reproduction simply adds an additional fold to this model without seriously calling it into question. It does this by stating that the root of identity is not to be located in the species but instead in the genetic individual.
13. On this difference, see Maturana and Varela, *Tree of Knowledge,* 87–90. To be specific, what regeneration requires are totipotent cells, be they

stem cells or dedifferentiated somatic cells. On this process, see especially Momoko Ikeuchi, Yoichi Ogawa, Akira Iwase, and Keiko Sugimoto, "Plant Regeneration: Cellular Origins and Molecular Mechanisms," *Development* 143, no. 9 (2016):1442–51.

14. To be precise, sex requires the merging of two gametes in the formation of a zygote. Just as with binary fission, it is difficult to identify, physiologically speaking, an event of death. But neither is it straightforwardly clear that such a transformation does not entail death, from the perspective of phenomenological or structural individuality.
15. One can, of course, grow apple trees from seed. In this case, the "child" will not be genetically identical to the parent, because apples are extreme heterozygotes, meaning that the seeds contain great genetic variation compared to the parent that produces them—every seed, in fact, even with respect to the other seeds of the same tree. On the importance of the germ line for plants, see Robert Lanfear, "Do Plants Have a Segregated Germline?," *PLoS Biology* 16, no. 5 (2018): e2005439.
16. For the story of Newton's tree and its afterlives, see Merlin Sheldrake, *Entangled Life* (New York: Random House, 2021), 218–22.

12. Biological Immortality III

1. For another example of an extremely long-lived plant, see *Azorella compacta,* specimens of which are believed to exceed three thousand years of age, as well as *Welwitschia.* Instances of lichens—hybrid organisms consisting of fungi and a microscopic photosymbiont (either an alga or a cyanobacterium)—have been known to live for upward of nine thousand years.
2. R. M. Lanner and K. F. Connor, "Does Bristlecone Pine Senesce?," *Experimental Gerontology* 36 (2001): 675–85, 675.
3. Lanner and Connor, "Does Bristlecone Pine Senesce?," 683.
4. There are of course many organisms, such as the semelparous octopus, whose life cycles include death, not due to senescence but as a naturalized stage of their development.
5. David A. Sinclair with Matthew D. LaPlante, *Lifespan: Why We Age—And Why We Don't Have To* (New York: Atria, 2019), 41. Here Sinclair lays out his "Information Theory of Aging": "Epigenetic noise causes the same kind of chaos. It is driven in large part by highly disruptive insults to the cell, such as broken DNA, as it was in the original survival circuit of *M. superstes* and in the old yeast cells that lost their fertility. And this, according to the Information Theory of Aging, is why we age. It's why

our hair grays. It's why our skin wrinkles. It's why our joints begin to ache. Moreover, it's why each one of the hallmarks of aging occurs, from stem cell exhaustion and cellular senescence to mitochondrial dysfunction and rapid telomere shortening" (38). For another recent work dedicated to lengthening life-span, see Andrew Steele, *Ageless* (New York: Doubleday, 2021).

6. Howard Thomas, "Senescence, Ageing, and Death of the Whole Plant," *New Phytologist* 197 (2013): 696–711, 697.
7. Thomas, "Senescence," 698.
8. Thomas, "Senescence," 707.
9. As to whether senescence is "preprogrammed" or not, Thomas is circumspect: "That a process proceeds towards a final state in a predictable fashion is not, by itself, evidence that it is genetically programmed; it may simply be that the system is built in a particular way and so always fails in more or less the same manner." Thomas, "Senescence," 699. To further disentangle the often conflated processes of aging, senescence, and death, Thomas proposes differentiating between gerontology (the study of aging and senescence) and acherontology (the study of "causes and symptoms of dying as distinct from ageing," such as apoptosis). Nevertheless, Thomas questions whether plants could serve as a model of comparison for animal aging: "It is questionable whether the mechanisms of ageing in plants are related other than very distantly to those of animals, or even whether plants undergo ageing in any gerontologically-recognisable sense. . . . This conclusion arises from consideration of unique structural, functional, and genetic characteristics that equip plants to avoid, resist or exploit the inevitability of ageing." Thomas, "Do Green Plants Age," 147.
10. See Daniel E. Martínez, "Mortality Patterns Suggest Lack of Senescence in Hydra," *Experimental Gerontology* 33, no. 3 (1998): 217–25, 217.
11. Martínez, "Mortality Patterns," 217–18.
12. Martínez, "Mortality Patterns," 218. Other examples of animals that rejuvenate include *Botryllus schlosseri* and *Schmidtea mediterranea.*
13. Martínez, "Mortality Patterns," 224.
14. Martínez, "Mortality Patterns," 218. As Birnbaum and Sánchez Alvarado note, "The long life span observed in many invertebrates is associated with the lifelong maintenance and regulation of stem cells in somatic tissues. The progeny of these adult stem cells are capable of replacing dying differentiated cells, allowing such organisms to effectively escape death." Kenneth D. Birnbaum and Alejandro Sánchez Alvarado, "Slicing across Kingdoms: Regeneration in Plants and Animals," *Cell* 132, no. 4 (2008): 697–710, 698.

15. See S. Piraino, F. Boero, B. Aeschbach, and V. Schmid, "Reversing the Life Cycle: Medusae Transforming into Polyps and Cell Transdifferentiation in *Turritopsis nutricula* (Cnidaria, Hydrozoa)," *Biological Bulletin* 190, no. 3 (1996): 302–12.
16. S. Piraino, D. De Vito, J. Schmich, J. Bouillon, and F. Boero, "Reverse Development in Cnidaria," *Canadian Journal of Zoology* 82 (2004): 1748–54, 1748. On the jellyfish, see Rich's wonderful piece on Japanese scientist Shin Kubota: Nathaniel Rich, "Can a Jellyfish Unlock the Secret of Immortality?," *New York Times,* November 28, 2012, https://www.nytimes.com/2012/12/02/magazine/can-a-jellyfish-unlock-the-secret-of-immortality.html.
17. As Alvarado and Yamanaka remark, "Complete adult organisms can revert in their entirety to a larval state through reprogamming mechanisms that remain to be elucidated." Alejandro Sánchez Alvarado and Shinya Yamanaka, "Rethinking Differentiation: Stem Cells, Regeneration, and Plasticity," *Cell* 157 (2014): 110–19, 114.
18. Piraino et al., "Reversing the Life Cycle," 302.
19. It is worth noting that while *T. dohrnii* do this naturally—or, at any event, with only slight manipulation within laboratory settings—it is suspected that such cellular dedifferentiation and reprogramming is a latent capacity much more broadly possessed than is commonly thought. Even in the day of Weismann (the father of the theory of germ line–soma split), there were already data to counter his assertion that somatic cells' fates were irreversible. It is on the potential of the yet unknown limits of cellular rejuvenation and pluripotency that much of the hope for antiaging or desenescence technologies rests today. As Alvarado and Yamanaka, "Rethinking Differentiation," put it, "The fact that it is possible to reprogram cells by either perturbing their environment or changing their genomic output artificially indicates that the plasticity of the differentiated state may not be restricted to simple animals and that programs for dedifferentiation may be more prevalent and much more broadly distributed among all animals (including humans) than most care to contemplate at the present time" (115). On this, see also Sinclair with LaPlante, *Lifespan,* 158–75.
20. Piraino et al., "Reversing the Life Cycle," 302. The authors give the following as the reason for this spectacular capacity: "The transformation process of *T. nutricula* medusae can be explained, at the cellular level, by two alternative, but not mutually exclusive, hypotheses. First, all differentiated somatic cells of the medusa degenerate, and the production of polyp cells is initiated by a set of undifferentiated reserve cells that were

not irreversibly committed. Second, differentiated cells of the medusa might transdifferentiate [or dedifferentiate] to produce the requisite new cell types" (303).

21. Piraino et al., "Reverse Development in Cnidaria," 1750. For an exploration of the notion of self and modes of cognition of such transitions, especially as they pertain to hypothetical technologies of life extension, see Levin, "Life, Death, and Self." As he puts it, "Such transformations are far more common than most assume; indeed, given the coming advances of regenerative medicine of the brain, and the increasing sophistication of brain-machine interfaces, you (or your children) are likely to someday find out, first hand, what it's like to undergo a signification modification of the biological substrate underlying your mind. This is the province of a newly emerging, interdisciplinary subfield at the intersections of cognitive science, regenerative biology, synthetic bioengineering, and neuroscience beyond neurons" (1).
22. Piraino et al., "Reverse Development in Cnidaria," 1752.

13. Metabolism and the Meaning of Life

1. Weber and Varela, "Life after Kant," 112. Throughout this chapter, I draw heavily on this essay.
2. On the importance of Kant within the history of biological conceptions of life, see Weber and Varela, "Life after Kant"; and Hui, *Recursivity and Contingency.* On self-organization as a property of both living and nonliving systems, see Carlos Gershenson, Vito Trianni, Justin Werfel, and Hiroki Sayama, "Self-Organization and Artificial Life," *Artificial Life* 26 (2020): 391–408.
3. Weber and Varela, "Life after Kant," 97.
4. Weber and Varela, "Life after Kant," 101.
5. See F. G. Varela, H. R. Maturana, and R. Uribe, "Autopoiesis: The Organization of Living Systems, Its Characterization and a Model," *Biosystems* 5 (1974): 187–96; Humberto Maturana and Francisco Varela, *Autopoiesis and Cognition* (Dortrecht: Reidel, 1980); and Jesper Hoffmeyer, *Signs of Meaning in the Universe* (Bloomington: Indiana University Press, 1997).
6. See Moreno and Mossio, *Biological Autonomy;* Thompson, *Mind in Life;* and Damasio, *Strange Order.* For a recognition of the connection between homeostasis, autopoiesis, and Jonas's work, see Kingson Man and Antonio Damasio, "Homeostasis and Soft Robotics in the Design of Feeling Machines," *Nature Machine Intelligence* 1 (2019): 446–52.

7. Lynn Margulis, *Slanted Truths* (New York: Copernicus, 1997), 267–69. See also Lynn Margulis and Dorion Sagan, *Acquiring Genomes: A Theory of the Origin of Species* (New York: Basic Books, 2003), 40.
8. For more on the significance of the economic sense of metabolism for the thinking of life, see Adam R. Rosenthal, "From Metabolism to (Gift) Exchange: On the Propensity to Propensity," *Derrida Today* 17, no. 3 (2024).
9. There are, of course, exceptions. Cryptobiosis names a physiological state that certain organisms, such as tardigrades, nematodes, and plant seeds, can enter in which all, or nearly all, metabolic activity is suspended, but without leading to death or the breakdown of anatomical structure. For an overview of cryptobiosis, see James Clegg, "Cryptobiosis—A Peculiar State of Biological Organization," *Comparative Biochemistry and Psychology Part B: Biochemistry and Molecular Biology* 128, no. 4 (2001): 613–24. Clegg provocatively claims that "we may conclude that there are three states of biological organization: alive; dead; and cryptobiotic" (615).
10. On the differences between living and nonliving dissipative systems, see Alicia Juarrero, "What Does the Closure of Context-Sensitive Constraints Mean for Determinism, Autonomy, Self-Determination, and Agency?," *Progress in Biophysics and Molecular Biology* 119 (2015): 510–21. On the generative role of encoded information for life, see Sara Walker and Paul Davies, "The Algorithmic Origins of Life," *Journal of the Royal Society Interface* 10 (2012): 1–9. On the distinction with autocatalytic systems, see Thompson, *Mind in Life,* 104–5. Rather than a firm distinction between some living and some nonliving realms, we can think of the need or mortality that such dependency inscribes into the structure of the living, evolutionarily speaking, as the cause of the attributes we commonly associate with life. In other words, were such need, dependency, or mortality written into inorganic constructs, then perhaps the latter would take on more lifelike attributes. For such a proposal, see Man and Damasio, "Homeostasis and Soft Robotics."
11. On this difference, see Julian Kiverstein and Erik Rietveld, "Reconceiving Representation-Hungry Cognition: An Ecological-Enactive Proposal," *Adaptive Behavior* 26, no. 4 (2018): 147–63.
12. Kant, *Critique of the Power of Judgment,* 243.
13. Kant, *Critique of the Power of Judgment,* 245.
14. Kant, *Critique of the Power of Judgment,* 243. Each of Kant's three modes of self-organization (or self-accounting) highlights a distinct logic of supplementarity at work in the living. Collectively, they identify the living

with a general capacity for recursive prostheticization: the organized and self-organized being is what it is only through the mutual interactions of fundamentally complex relations that are not simply *in* space and time, but that *produce* space and time as their very work. Reproduction, production, and self-maintenance establish the being of the self-organized being only through what it is not and what it might not be, as a thoroughly plastic—which is to say divisible—whole. An open question is whether something like metabolism, now understood in the most broad, formal sense of the term, as *precarious exchange,* is a necessary precondition for all self-organization worthy of the name. Put otherwise, it seems that the conceit of autopoietic theory is that metabolism imbues self-organization with its life-like characteristics: the capacity for experience, or affection.

15. Weber and Varela, "Life after Kant," 116.
16. For an overview of the history of autopoietic theory and its importance for Margulis and Lovelock's thinking of Gaia, see Bruce Clarke, *Gaian Systems: Lynn Margulis, Neocybernetics, and the End of the Anthropocene* (Minneapolis: University of Minnesota Press, 2020).
17. At another level of organization, this is also the function of the immune system. On the implications for selfhood of immunitary processes, see Thomas Pradeau and Edgardo D. Carosella, "The Self Model and the Concept of Biological Identity in Immunology," *Biology and Philosophy* 21 (2006): 235–52; and Thomas Pradeau, *The Limits of the Self: Immunology and Biological Identity,* trans. Elizabeth Vitanza (Oxford: Oxford University Press, 2012).
18. On the necessity of thinking autopoiesis together with cognition, see Pier Luigi Luisi, "Autopoiesis: A Review and a Reappraisal," *Naturwissenschaften* 90 (2003): 49–59. On unminded and unconscious forms of cognition, see Damasio, *Strange Order;* and Marder, *Plant-Thinking.* It is on the possibility of such a distinction that one can speak of the world of plants, and by extension all life, as an expression of their potentialities. Where there is actionable information, we could say, there is world. In this sense, even if theories of abiogenesis must suppose the (historical) existence of compartmentalization prior to the metabolic activity that perpetuates them, it is only after this spatial difference becomes taken up by the regulative feedback loops of the autopoietic system that it gains its sense as border. Thompson is helpful here: "According to the view I have been proposing, autopoiesis plus adaptivity entails sense-making, which is cognition in its minimal biological form." Thompson, *Mind in Life,* 159. For a more extreme version of this

thought, see the cellular basis of consciousness theory in Reber, *First Minds*. Finally, for a helpful account of the related notion of basal cognition, see Michael Levin and Daniel Dennett, "Cognition All the Way Down," *Aeon*, October 13, 2020, https://aeon.co/essays/how-to-understand-cells-tissues-and-organisms-as-agents-with-agendas.

19. On this, see Moreno and Mossio, *Biological Autonomy*.
20. Jonas, *Phenomenon of Life*, xxiii.
21. Jonas, *Phenomenon of Life*, xxiii. As Weber and Varela write, "Jonas is unique in demanding that the analysis be carried to the minimal form of life, to its very origins, and to where it joins the autopoiesis account. It is from this minimal understanding that the qualities of autonomy and purpose can eventually be echoed in the multicellular organisms endowed with a nervous system." Weber and Varela, "Life after Kant," 112. It should be noted that although Jonas does try to locate forms of individuality and autonomy in all life, his views of human/nonhuman animal difference are often highly problematic. See Jonas, *Phenomenon of Life*; and Hans Jonas, *Mortality and Morality* (Evanston, Ill.: Northwestern University Press, 1996), 75–86.
22. See Georges Canguilhem, "Le concept et la vie," *Revue philosophique de Louvain* 64, no. 82 (1966): 193–223, 193.
23. On life as planetary phenomenon, see Owen, "Life as a Planetary Phenomenon"; and Lovelock and Margulis, "Atmospheric Homeostasis."
24. Donna J. Haraway, *When Species Meet* (Minneapolis: University of Minnesota Press, 2007), 32–33. See also Donna J. Haraway, *Staying with the Trouble: Making Kin in the Chthulucene* (Durham, N.C.: Duke University Press, 2016), 58–98. In the latter text, regarding "sympoiesis," she further specifies: "As long as autopoiesis does not mean self-sufficient 'self-making,' autopoiesis and sympoiesis, foregrounding and backgrounding different aspects of systemic complexity, are in generative friction, or generative enfolding, rather than opposition" (61). For another important discussion of autopoiesis, see Cary Wolfe, "In the Shadow of Wittgenstein's Lion," in *Animal Rites* (Chicago: University of Chicago Press, 2003).
25. For a typical quantitative approach to the organism in terms of degrees of conflict, see Queller and Strassmann, "Beyond Society": "The most salient feature of organisms is adaptation, the seeming goal-directedness that makes organisms different from merely physical entities. . . . Every organism has numerous adaptations, all directed to the growth, development and reproduction of that organism (and sometimes its kin). We suggest that the essence of organismality lies in the shared purpose; the parts work together for the integrated whole, with high cooperation and

very low conflict. Specifically, the organism is the largest unit of near-unanimous design; the qualifying 'near' is required because some conflicts, like meiotic drive, probably remain in all organisms. All organism concepts emphasize function and integration; ours differs in stressing unanimity, and in ignoring other traits. Our definition is a social one; the organism is simply a unit with high cooperation and very low conflict among its parts. That is, the organism has adaptations and it is not much disrupted by adaptations at lower levels" (3144).

26. On this, see Díaz-Muñoz et al., "Contextual Organismality." Nor is it as simple a matter as identifying consciousness as the single, unifying, and undeniable principle of "organism" or "self." This is not the case for at least two reasons: first, because it is a notoriously difficult quality to identify, and second, because the ultimate privilege bestowed on consciousness (as favored by proponents of brain uploading, for instance) denies the extreme and undeniable complexity of organization, which defies reduction to any single quality. Who is to say that "I" have one consciousness? Is not the effect of consciousness precisely to present the vision of indivisible unity? But here is the point: one can always decide to value consciousness as the final resting place of the soul or presence. Fine. But that decision can never be justified on objective grounds. Rather than the ultimate solution, it is merely one more solution, all the more violent for its foreclosure of every difference. The recourse to consciousness, in sum, solves nothing but misses much. For the most direct possible rebuttal to this position—that is, the prospect of multiple consciousnesses housed in a single body—see Peter Godfrey-Smith, *Other Minds* (New York: Farrar, Straus & Giroux, 2017); and Peter Godfrey-Smith, *Metazoa: Animal Life and the Birth of the Mind* (New York: Farrar, Straus & Giroux, 2020).
27. As Weber and Varela put it: "We think, in contrast [to efforts to naturalize teleology through reduction to mechanism], that an integration of teleological descriptions can only be possible by taking them seriously: by accepting that organisms are subjects having purposes according to values encountered in the making of their living. This means clearly to reintroduce value and subjectivity as indispensable organic phenomena, a theory of the organism as the dynamics of establishing an identity and, hence, as a process of creating a materially embodied, individual perspective." Weber and Varela, "Life after Kant," 102.
28. For another important response to Haraway and Gilbert, see Derek Woods, "Prosthetic Symbiosis," *CR: The New Centennial Review* 22, no. 1 (2022): 157–86.
29. Jonas, *Phenomenon of Life*, 99–107.

30. Jonas, "Biological Foundations," 233.
31. Jonas, "Biological Foundations," 232.
32. Jonas, "Biological Foundations," 232.
33. Jonas, "Biological Foundations," 232.
34. Jonas, "Biological Foundations," 233.
35. Jonas, "Biological Foundations," 235.
36. Jonas, "Biological Foundations," 237.
37. Jonas, "Biological Foundations," 239.
38. Jonas, "Biological Foundations," 239.
39. Jonas, "Biological Foundations," 241.
40. Jonas, "Biological Foundations," 242.
41. Jonas, "Biological Foundations," 242–43.
42. Jonas, "Biological Foundations," 243. "The centeredness of animal organization brings out the 'indivisibility' which lies in the literal meaning of 'individual.' And it is the order of individuation that marks the ontological difference between animal and plant. . . . To be an individual means not-to-be-integrated with the world, and the less so, the more it is an individual. Individuality implies discontinuity" (250–51).

Part IV

1. Bostrom, "Ethical Issues in Advanced Artificial Intelligence," 282. On whole brain emulation, or mind uploading, see, variously: Hans Moravec, *Mind Children* (Cambridge, Mass.: Harvard University Press, 1990); Gregory S. Paul and Earl Cox, *Beyond Humanity* (Rockland, Mass.: Charles River Media, 1996); and Natasha Vita-More and Max More, eds., *The Transhumanist Reader* (Hoboken, N.J.: Wiley, 2013).
2. This is a bare minimum or maximum, for, according to many transhumanists, liberation from the body entails an expansion, and not a reduction, in human capacities. If I emphasize the minimum, it is to indicate that even in this maximalist interpretation, what are revealed are the minimal conditions for self-translation. For simplicity's sake, I will adopt Bostrom's figure of age reversal, by which is meant senescence reversal and/or stabilization.
3. For an important recent counterapproach to Bostrom's posthumanism that differs from my own, see Claire Colebrook and Jamie Weinstein, eds., *Posthumous Life: Theorizing beyond the Posthuman* (New York: Columbia University Press, 2017).
4. Doyle observes how the neo-Darwinian "conflation of life and information" provides a theoretical basis for uploading. Richard Doyle, *Wetwares:*

Experiments in Postvital Living (Minneapolis: University of Minnesota Press, 2004), 74.

5. As Chalmers puts it, such an elimination of the biological substrate is not only possible but necessary if we are to compete with future AI systems: "In the long run, if we are to match the speed and capacity of nonbiological systems, we will probably have to dispense with our biological core entirely. This might happen through a gradual process through which parts of our brain are replaced over time, or it happen through a process of scanning our brains and loading the result into a computer, and then enhancing the resulting processes. Either way, the result is likely to be an enhanced nonbiological system, most likely a computational system." David Chalmers, "Singularity," *Journal of Consciousness Studies* 17, no. 9–10 (2010): 7–65, 41–42.

14. Age Reversal

1. For a brief exploration of these problems, see the chapter on "Immortality" in Harris, *Enhancing Evolution.* In Parfit's case, he makes the rather radical claim that personal identity does not matter. Nevertheless, in concluding that what does matter is Relation R, he constructs a system within which robust psychological continuity is not only possible but also the true foundation for ipseity, now marked by the self-presence of its desires to itself.
2. See de Grey and Rae, *Ending Aging,* for one such list.

15. Whole Brain Emulation

1. On the conceits of such a conception, see Michael Rescorla, "The Computational Theory of Mind," *Stanford Encyclopedia of Philosophy* (2020), https://plato.stanford.edu/entries/computational-mind/.
2. Anders Sandberg and Nick Bostrom, *Whole Brain Emulation: A Roadmap,* Technical Report 2008-3, Future of Humanity Institute, Oxford University (2008), 7, https://www.fhi.ox.ac.uk/brain-emulation-roadmap-report.pdf.
3. In the strongest version of brain uploading, there is not only simulation but also emulation, down to the neuron and even the atom—all of which of course begs the question as to whether biological life can be reproduced in substrate-independent fashion by way of simulated components that do not actually bear the chemical and physical properties of organic chemistry.

4. For a few typical examples of this position, see Pigliucci's response to Chalmers: Massimo Pigliucci, "Mind Uploading: A Philosophical Counter-analysis," in *Intelligence Unbound,* ed. Russel Blackford and Damien Broderick (New York: Wiley, 2014), 119–30; and Reber, *First Minds.*
5. For a helpful overview of the transhumanist movement and its problematic aspects, see Francesco Paolo Adorno, *The Transhumanist Movement* (London: Palgrave Macmillan, 2021).
6. Marvin Minsky, "Will Robots Inherit the Earth?," *Scientific American* 271, no. 4 (1994): 108–13, 113.
7. As Chalmers puts it, "Within the next century, however, we may construct reasonably accurate simulations of human brains and behavior. Sometime after that, we might have plausible simulations of a whole human society. Eventually we might simulate a solar system or even a universe, from the level of atoms to the level of the cosmos. In such a system, there will be bits corresponding to every entity in the universe being simulated. Once we have fine-grained simulations of all the activity in a human brain, we'll have to take seriously the idea that the simulated brains are themselves conscious and intelligent." David Chalmers, *Reality+* (New York: Norton, 2022), 22–23.
8. Chalmers, "Singularity," 42.
9. Chalmers, "Singularity," 42.
10. Chalmers, "Singularity," 42.
11. This passage from Chalmers is telling: "Once we accept that gradual uploading over a period of years might preserve identity, the obvious next step is to speed up the process. Suppose that Dave's brain is gradually uploaded over a period of hours, with neurons replaced one at a time by functionally isomorphic silicon circuits. Will Dave survive this process? It is hard to see why a period of hours should be different in principle from a period of years, so it is natural to hold that Dave will survive. . . . We can then imagine speeding up the process from hours to minutes. The issues here do not seem different in principle. One might then speed up to seconds. At a certain point, one will arguably start replacing large enough chunks of the brain from moment to moment that the case for continuity of consciousness between moments is not as secure as it is above. Still, once we grant that uploading over a period of minutes preserves identity, it is at least hard to see why uploading over a period of seconds should not. As we upload faster and faster, the limit point is instant destructive uploading, where the whole brain is replaced at once.

Perhaps this limit point is different from everything that came before it, but this is at least unobvious. We might formulate this as an argument for the optimistic view of destructive uploading. Here it is to be understood that both the gradual uploading and instant uploading are destructive in that they destroy the original brain. . . . It is hard to see why there is a difference in principle between uploading over a period of seconds and doing so instantly. As before, this argument parallels a corresponding argument about teletransportation." Chalmers, "Singularity," 53–54.

12. On teletransportation, see Douglas Hofstadter and Daniel Dennett, *The Mind's I* (Toronto: Bantam, 1982); Parfit, *Reasons and Persons;* and, for a helpful synopsis, Charlie Huenemann, "If I Teleport from Mars, Does the Original Me Get Destroyed?," *Aeon,* August 1, 2017, https://aeon.co/ideas/if-i-teleport-from-mars-does-the-original-me-get-destroyed.
13. Chalmers, "Singularity," 49.
14. Pigliucci, "Mind Uploading."
15. Michel de Montaigne, "De la gloire," in *Les Essais. Livre II* (Paris: Presses universitaires de France, 1999), 618–30, 618. The translation is my own.
16. As Sandberg and Bostrom put it, "If emulation of particular brains is possible and affordable, and if concerns about individual identity can be met, such emulation would enable back-up copies and 'digital immortality.' " Sandberg and Bostrom, *Whole Brain Emulation,* 5. They also give the following description of the criteria for "personal identity emulation": "The emulation is correctly described as a continuation of the original mind; either as numerically the same person, or as a surviving continuer thereof" (11). Pigliucci puts the problem as follows: "And there lies the rub for all discussions about personal identity: we make a certain sense of the concept because human beings are a particular type of biological species (biology again!). If we were a very different type of being, we would have a very different concept of personal identity, or perhaps even no such concept at all. There is, in other words, no *metaphysical fact* of the matter about personal identity—which perhaps helps explain both Chalmers's own lack of commitment on the subject and the fact that neo-scholastic philosophers simply cannot agree on it." Pigliucci, "Mind Uploading," 127.
17. Though teletransportation is the classic example that is given for uploading, is not its true meaning that of time travel? The compact between present and future self, which is uploading's secret sense, is nothing less than the fiction, implicit in my current assurance, of having gone forward and then returned backward in time, to the present.

18. Chalmers, "Singularity," 46.
19. Chalmers, "Singularity," 53–55.
20. On this problem, see Jonas, "Biological Foundations," 239.
21. Chalmers, "Singularity," 53.
22. Does this not also imply that age reversal is itself a form of gradual uploading?
23. Parfit, *Reasons and Persons,* 280.
24. Parfit, *Reasons and Persons,* 289.
25. See also Derek Parfit, "Personal Identity," *Philosophical Review* 80, no. 1 (1971): 3–27.
26. Parfit, *Reasons and Persons,* 281.

16. Cloning and the Fictions of Immortality

1. Chalmers, "Singularity," 54.
2. For my earlier examination of the logic of the clone, see Adam R. Rosenthal, "Clon'd, or a Note on Propagation," *SubStance* 4, no. 1 (2019): 46–62.
3. Josie is ill because of the gene therapy she underwent to enhance her cognitive abilities. This normalized procedure was also the cause of her sister's earlier death, for which a first, and unsuccessful, "continuation" was made.
4. Kazuo Ishiguro, *Klara and the Sun: A Novel* (New York: Knopf, 2021), 103.
5. Ishiguro, *Klara,* 205.
6. Ishiguro, *Klara,* 206–8.
7. As a member of the faculty at Texas A&M University, I would be remiss not to mention the logic of continuation that regulates the economy of our mascot, Reveille, who is, at the moment of this writing and according to official count, the tenth in her line.
8. Michel Houellebecq, *La possibilité d'une île* (Paris: Flammarion, 2013), 325. Translations are my own.
9. Houellebecq, *Possibilité,* 359.
10. Houellebecq, *Possibilité,* 383.

Conclusion

1. Mary Shelley, "The Mortal Immortal," in *Mary Shelley: Collected Tales and Stories,* ed. Charles E. Robinson (Baltimore, Md.: Johns Hopkins University Press, 1976), 219–30, 219.

2. But is it a coincidence that the writing of Heidegger's existential analytic largely coincides with the rise of these very sciences in the first half of the twentieth century? In other words, cannot the historicity of the *sein-zum-Tode,* however irrefutable it is today and however universal it may seem, itself be circumscribed within the same epoch as the becoming physiological of life?
3. Shelley, "Mortal Immortal," 229.

INDEX

(continued from page ii)

58 *Individuation in Light of Notions of Form and Information*
Volume II. *Supplemental Texts*
Gilbert Simondon

57 *Individuation in Light of Notions of Form and Information*
Gilbert Simondon

56 *Thinking Plant Animal Human: Encounters with Communities of Difference*
David Wood

55 *The Elements of Foucault*
Gregg Lambert

54 *Postcinematic Vision: The Coevolution of Moving-Image Media and the Spectator*
Roger F. Cook

53 *Bleak Joys: Aesthetics of Ecology and Impossibility*
Matthew Fuller and Olga Goriunova

52 *Variations on Media Thinking*
Siegfried Zielinski

51 *Aesthesis and Perceptronium: On the Entanglement of Sensation, Cognition, and Matter*
Alexander Wilson

50 *Anthropocene Poetics: Deep Time, Sacrifice Zones, and Extinction*
David Farrier

49 *Metaphysical Experiments: Physics and the Invention of the Universe*
Bjørn Ekeberg

48 *Dialogues on the Human Ape*
Laurent Dubreuil and Sue Savage-Rumbaugh

47 *Elements of a Philosophy of Technology: On the Evolutionary History of Culture*
Ernst Kapp

46 *Biology in the Grid: Graphic Design and the Envisioning of Life*
Phillip Thurtle

45 *Neurotechnology and the End of Finitude*
Michael Haworth

44 *Life: A Modern Invention*
Davide Tarizzo

43 *Bioaesthetics: Making Sense of Life in Science and the Arts*
Carsten Strathausen

42 *Creaturely Love: How Desire Makes Us More and Less Than Human*
Dominic Pettman

41 *Matters of Care: Speculative Ethics in More Than Human Worlds*
María Puig de la Bellacasa

40 *Of Sheep, Oranges, and Yeast: A Multispecies Impression*
Julian Yates

39 *Fuel: A Speculative Dictionary*
Karen Pinkus

38 *What Would Animals Say If We Asked the Right Questions?*
Vinciane Despret

37 *Manifestly Haraway*
Donna J. Haraway

36 *Neofinalism*
Raymond Ruyer

35 *Inanimation: Theories of Inorganic Life*
David Wills

34 *All Thoughts Are Equal: Laruelle and Nonhuman Philosophy*
John Ó Maoilearca

33 *Necromedia*
Marcel O'Gorman

32 *The Intellective Space: Thinking beyond Cognition*
Laurent Dubreuil

31 *Laruelle: Against the Digital*
Alexander R. Galloway

30 *The Universe of Things: On Speculative Realism*
Steven Shaviro

29 *Neocybernetics and Narrative*
Bruce Clarke

28 *Cinders*
Jacques Derrida

27 *Hyperobjects: Philosophy and Ecology after the End of the World*
Timothy Morton

26 *Humanesis: Sound and Technological Posthumanism*
David Cecchetto

25 *Artist Animal*
Steve Baker

24 *Without Offending Humans: A Critique of Animal Rights*
Élisabeth de Fontenay

23 *Vampyroteuthis Infernalis: A Treatise, with a Report by the Institut Scientifique de Recherche Paranaturaliste*
Vilém Flusser and Louis Bec

22 *Body Drift: Butler, Hayles, Haraway*
Arthur Kroker

21 *HumAnimal: Race, Law, Language*
Kalpana Rahita Seshadri

20 *Alien Phenomenology, or What It's Like to Be a Thing*
Ian Bogost

19 *CIFERAE: A Bestiary in Five Fingers*
Tom Tyler

18 *Improper Life: Technology and Biopolitics from Heidegger to Agamben*
Timothy C. Campbell

17 *Surface Encounters: Thinking with Animals and Art*
Ron Broglio

16 *Against Ecological Sovereignty: Ethics, Biopolitics, and Saving the Natural World*
Mick Smith

15 *Animal Stories: Narrating across Species Lines*
Susan McHugh

14 *Human Error: Species-Being and Media Machines*
Dominic Pettman

13 *Junkware*
Thierry Bardini

12 *A Foray into the Worlds of Animals and Humans, with A Theory of Meaning*
Jakob von Uexküll

11 *Insect Media: An Archaeology of Animals and Technology*
Jussi Parikka

10 *Cosmopolitics II*
Isabelle Stengers

9 *Cosmopolitics I*
Isabelle Stengers

8 *What Is Posthumanism?*
Cary Wolfe

7 *Political Affect: Connecting the Social and the Somatic*
John Protevi

6 *Animal Capital: Rendering Life in Biopolitical Times*
Nicole Shukin

5 *Dorsality: Thinking Back through Technology and Politics*
David Wills

4 *Bíos: Biopolitics and Philosophy*
Roberto Esposito

3 *When Species Meet*
Donna J. Haraway

2 *The Poetics of DNA*
Judith Roof

1 *The Parasite*
Michel Serres

Adam R. Rosenthal is associate professor of global languages and cultures at Texas A&M University. He is author of *Poetics and the Gift: Reading Poetry from Homer to Derrida.*

David Wills is professor of French studies at Brown University and author of *Prosthesis,* also published by the University of Minnesota Press.

Printed and bound by CPI Group (UK) Ltd, Croydon, CR0 4YY

11/09/2024

14554939-0001